T0230503

Springer INdAM Series

Volume 15

More information about this series at http://www.springer.com/series/10283

Fabio Ancona • Piermarco Cannarsa •
Christopher Jones • Alessandro Portaluri
Editors

Mathematical Paradigms of Climate Science

Springer

Editors
Fabio Ancona
Dipartimento di Matematica
Università degli Studi di Padova
Padova, Italy

Piermarco Cannarsa
Dipartimento di Matematica
Università di Roma - Tor Vergata
Roma, Italy

Christopher Jones
Department of Mathematics
University of North Carolina
Chapel Hill
North Carolina, USA

Alessandro Portaluri
Scienze Agrarie, Forestali e Alimentari
Università degli Studi di Torino
Torino, Italy

ISSN 2281-518X   ISSN 2281-5198   (electronic)
Springer INdAM Series
ISBN 978-3-319-81815-3   ISBN 978-3-319-39092-5   (eBook)
DOI 10.1007/978-3-319-39092-5

Preface

We are living in a changing climate. We now know that the changes to come will require us to adjust our lifestyles and likely adapt to a very different kind of planet from the one humans have enjoyed since we have been around as a species. We will have to take this seriously as individuals. The question posed here is whether we should play a role as mathematicians by bringing our expertise to bear on the issues that are emerging in climate science.

A workshop on the subject was held in Rome during the summer of 2013. Mathematicians of varying stripes came together with some, mathematically inclined, climate scientists. We heard a variety of lectures, ranging from ones that presented hard problems posed by climate to mathematical lectures filled with new ideas and techniques that show promise for application in this area. As a group, we spent considerable time brainstorming through fundamental issues where mathematical work could make a difference.

The potential for mathematical concepts, investigations, and techniques is enormous in almost all areas of climate science, and the need is great. Nevertheless, it would be hard to give a definition of the area of *mathematical climate science* at this point in time. Massive computational models of the Earth system lie at the center of much of the work in climate science. These are mathematical models built around the dynamical core of circulation models based on the classical equations of fluid mechanics. With the notable exception of numerical analysts devoted to improving algorithms, most mathematicians find less comprehensive models more conducive to the development of ideas and techniques. But this is also true from the climate scientists' viewpoint. While the *big* models inevitably provide greater detail than conceptual models, it can be hard to see relations between different physical effects because there is so much going on. Smaller conceptual models strip away much of the extraneous information and allow focus on a particular phenomenon.

In many areas, however, it is not clear what models we should be using. In particular, this is true in biogeochemistry. The interaction of chemicals and biota in the ocean plays a critical role in the response of the ocean to increased CO_2 in the atmosphere. But the essential components of this process are not well understood, and the general area of ocean biogeochemistry poses enormous modeling challenges. There are common features and issues in the proposed models, and mathematics could play a significant role in clarifying and defining this area by guiding research toward better models through systematic reduction and laying the groundwork for the analysis of typical models. The first paper in this volume addresses this area and introduces some of the key problems.

Two mathematical areas were isolated through our discussions as providing common fertile ground for climate science and applied mathematics: *stochastic differential equations* and *control theory*. To understand why these areas emerged requires some explanation of the complementary roles of data and models in climate science. By data, we are referring to the plethora of observational measurements we now have of our climate. It may seem that we have enough data to actually estimate the climatic conditions of our Earth. But this is not true owing to the Earth's immensity and the complexity of all the different processes that impact its climate. Indeed, one way to look at the need for models is to interpolate between data points in order to render a full picture. The use of a model for this interpolation guarantees that it is achieved in accordance with the physical laws that underlie the model. This is to look at the situation from the data viewpoint. From the modeler's vantage point, the observational data are used as a constraint on the model. No model is perfect, and its shortcomings will cause it to produce inaccurate estimates or predictions. This problem can be mitigated by using observational data to force the model closer to reality.

Viewed either way, there is an intrinsic balance between models and data that lies at the heart of climate investigations. The mathematical topic that underpins this balancing act is called *data assimilation (DA)*. The topic is posed in terms of stochastic DEs, and this is manifest in the papers by Carrassi and Vannitsem as well as that by McDougall and Jones. Much of DA in the context of applied mathematics has its origins in control theory, and this emerges clearly in the paper by Blum and Auroux.

The papers by Alabau et al. and by Porretta and Zuazua show what advances are being made in control theory that promise application in areas of climate. They are not posed directly in terms of climate problems, but show the kinds of question that are being addressed at an advanced level of mathematical control theory. The challenge for us is to see how to build a bridge between this work and critical climate problems.

There are two areas where mathematics has made strong inroads in climate science: *paleoclimate* and *ocean dynamics*. Two papers in each area are featured in this volume. Gairing et al. discuss the issues of Lévy noise in paleoclimate time series, while Crucifix and Mitsui address stochastic effects in glacial cycles.

The Navier-Stokes equations form the basis of the dynamical core of all climate models. The influence of stochasticity on these equations is an important issue that is covered in the paper by Flandoli. The critical effect of convection in the ocean and vertical overturning form the focus of the paper by Paparella.

The backstory to this workshop is a group of researchers gathering together to build an area we might call the *Mathematics of Climate*. The goal of this volume is to show what might be possible and also to expose some of the challenges. In the future we shall need to determine the paradigmatic problems that will forge this area, and we hope that this volume will give some insights into how this endeavor might proceed.

Padova, Italy   Fabio Ancona
Roma, Italy   Piermarco Cannarsa
Chapel Hill, USA   Christopher Jones
Torino, Italy   Alessandro Portaluri
June 2016

Contents

Part IV   Data Assimilation

Part I
Core Climate Issues

A Mathematical Perspective on Microbial Processes in Earth's Biogeochemical Cycles

Christof Meile and Chris Jones

Abstract The quantitative analysis of biogeochemical cycles interfaces numerous scientific disciplines. These cycles are largely driven by the activity of microorganisms, which needs to be described in mathematical models. Numerous challenges arise from this: First, the challenge of scale, connecting microbes to global patterns across many orders of magnitude. Second, the mathematical treatment of complex natural processes require - aside from expert knowledge - the use of systematic and objective model reduction schemes. Third, models and diverse data need to be integrated efficiently. Here we discuss these three challenges, highlight promising avenues for the expansion of mathematical approaches in the study of Earth's biogeochemical cycles and propose concerted educational efforts fostering collaborations in mathematical and geoscience research to advance the field.

Keywords Mathematics of biogeochemical cycles • Microbial processes • Challenge of scale • Model reduction • Data integration

1 Introduction

Earth's climate is intrinsically linked to the composition of the atmosphere, which in turn depends on the cycling of elements, and the interactions between these cycles. The chemical forms in which elements exist determine their distribution between solid, aqueous or gaseous phases and hence between distinct reservoirs such as the ocean and the atmosphere. The transformation of elements between different chemical forms is largely driven by microbes [34], which exploit thermodynamic disequilibrium to support their metabolism [35], affecting the distribution and bioavailability of most elements at the Earth surface. Therefore, we need a clear understanding of microbial activities and how environmental conditions control

C. Meile (✉)
Department of Marine Sciences, The University of Georgia, Athens, GA 30602, USA
e-mail: cmeile@uga.edu

C. Jones
Department of Mathematics, University of North Carolina at Chapel Hill, Chapel Hill, NC 27599, USA

© Springer International Publishing Switzerland 2016
F. Ancona et al. (eds.), *Mathematical Paradigms of Climate Science*, Springer
INdAM Series 15, DOI 10.1007/978-3-319-39092-5_1

them, particularly when the external forcing (e.g., climate) is changing. This motivates the analysis of microbially-mediated processes in the study of larger scale environmental systems.

A key climate question is how the short term (0.1–1000 years) carbon (C) cycle responds to an increase in global surface temperatures. The short-term C cycle involves a balance between carbon fixation by photosynthetic organisms and organic carbon mineralization (conversion to CO_2) by heterotrophic microbes. Carbon is distributed and stored in multiple pools, including terrestrial C (i.e., soil and vegetation), atmospheric C, and surface ocean C, which contain approximately 2500, 830 and 900 GtC, respectively [8]. In soils, most of this C is in an organic form, and thus susceptible to decomposition. An increase in temperature could stimulate greater mineralization of soil C to CO_2, which may then feed back to drive temperatures higher [38]. However, the mineralization of organic matter is a complex process, so the response to elevated temperatures may not be linear [41]. Carbon cycling is further coupled to cycles of nutrients [1] that are required for producing enzymes, nucleic acids, cell walls, etc. As a consequence, changes in one cycle can impact the transformation in another.

Falkowski et al. [14] point out that the interactions among oceanic, atmospheric and terrestrial carbon cycles are currently not well understood, yet their coupling will be crucial in fully assessing and predicting the reaction of the carbon cycle to anthropogenic radiative forcing. They advocate an integrated approach in which the various components of the carbon cycle are treated as part of a larger system. As discussed in Hajima et al. [19], Earth System Models provide a framework in which these various cycle components are included at a global scale. However, intermediate models are needed in which not only the isolated cycles can be analyzed, but also the feedbacks between them. Such intermediate scale models may integrate or be informed by models resolving yet smaller scales, e.g., the description of organic matter mineralization in soils resolving the interplay of transport and reactions. This hierarchy requires a mathematical framework that is transparent enough to lead to an understanding of the system functioning but also adaptable enough to serve as a module in a larger scale model.

In this brief communication, we first discuss that biogeochemical reactions depend on local substrate concentrations, which may vary over distances shorter than that resolved in a model applied at the macroscopic scale. This often leads to ad hoc simplified model representations, yet the implications of the choice of model formulation on predictions is rarely investigated. Secondly, as an example, we discuss efforts to quantify the fate of hydrocarbons in the aftermath of the large Deepwater Horizon spill in the Gulf of Mexico in 2010, and the associated mathematical tools and challenges. We outline approaches to the integration of data in models of complex microbially-driven biogeochemical cycles and discuss associated challenges. Finally, we summarize three general areas in which we see the potential for a significant impact of mathematical analyses in geosciences and the study of elemental cycles.

2 Microbially Mediated Biogeochemical Cycling: *The Challenge of Scale*

Microbial actors operate at a scale many orders of magnitude smaller than typical measurements of biogeochemical outcomes (e.g., CO_2 fluxes). Hence, the description of their effect on elemental cycling requires formulations that successfully capture the cumulative effect of processes occurring below the grid scale because models rarely resolve this local environment explicitly. Process rate expressions are commonly informed by expert knowledge (e.g., denitrification rates depending on substrate availability and inhibition by O_2), which are then parameterized based on the literature, laboratory experiments, or optimized to match observational data. While laboratory-determined kinetic constants have the advantage that they allow for controlled experimental conditions, they are often set up in well-mixed conditions. Under natural conditions, however, spatial variability below the scale resolved in models is common (e.g., pore scale variability in soils, suspended particles in aqueous solutions etc.). This has implications for the parameterization of reaction rates.

The basic model has the form of a advection-diffusion-reaction equation for the chemical species. We set $C = C(x, t)$ as the vector of concentrations depending on space (in principle 3-dimensional) and time. If there are n chemical substances involved, then $C = (c_1, c_2, \ldots, c_n)$ and the equation governing species distribution and interaction as they evolve in time will have the form:

$$\frac{\partial}{\partial t}(\varphi C) = \nabla \cdot (D \nabla C - uC) + R(C),$$

where φ is the porosity (1 in free fluid), $D = D(x)$ is the spatially varying effective (inhomogenous) diffusion which will depend on the nature of the chemical (molecular diffusion), ambient fluid (turbulent mixing) and/or the structure of the porous medium (hydrodynamic dispersion). Advection is described by the velocity field u, and the reactions between the various chemical species are contained in the term $R(C)$. Considering the case of a system with negligible advection at steady-state leads to

$$\nabla \cdot (D \nabla C) + R = 0.$$

A central problem is that the concentrations at the scale relevant for reactions (i.e., at the scale sensed by a microbe) are typically not experimentally accessible. A macroscopic formulation can be sought through homogenization [29], which essentially averages over volumes and replaces

$$\nabla \cdot \langle D \nabla C \rangle + R$$

with an expression based on volume averaged (and hence experimentally accessible) concentrations $\bar{C}$, such as

$$\nabla \cdot \left(D^* \nabla \bar{C}\right) + R^* = 0,$$

where D^* and $R^* = R^*(\bar{C}) \approx R(C)$ are effective parameterizations. In the case of homogeneous reactions, Meile and Tuncay [26] have used a Taylor series expansion of the reaction term, leading to

$$\langle R(C) \rangle \approx \varphi R\left(\bar{C}\right) + \frac{\alpha}{2} \left.\frac{d^2 R}{dC^2}\right|_{\bar{C}} \nabla \bar{C} \cdot \nabla \bar{C},$$

where α is a scale dependent phenomenological constant that depends on pore geometry, which they derived using numerical simulations at the pore scale. As a consequence, if one applies a rate parameter k, $R = f(k, C)$, determined in well-mixed laboratory conditions, to the field, this may introduce significant errors. This is the case when field observations provide average concentrations, or if the model resolution is insufficient to resolve concentrations at the relevant scale, so that $R(C) \neq \varphi R\left(\bar{C}\right)$, as demonstrated experimentally by Gramling et al. [17]. Macroscopic biogeochemical models deal with phenomena that cover a wide range of temporal and spatial scales. Microbial processes, however, are more complex than the above example of a homogeneous reaction, and may involve microbe-solid phase or microbe-microbe interactions, with largely unresolved implications for macroscopic approximations.

To tackle the challenge of scale, one may separate the timescales, so that slow processes (e.g., tied to geological processes) may be prescribed as environmental conditions/steady external forcing, intermediate ones be explicitly modeled (e.g., the dynamics in a food web), while fast processes (e.g., the adaptation of the microbial community) may be included in the model parameterization. Traditionally, this has been the most common approach, though accounting for variability in fast processes through a stochastic approach as noise may be more realistic [3].

Notably, Scheibe et al. [33] recently established a decision tree to guide the use of different multiscale simulation methods, and highlights the conditions they are most applicable to. If a full microscale model description is not feasible, then a coarse scale model can be well represented by using parameters derived from upscaling a high-resolution simulation in small subdomains, as done in the above example by Meile and Tuncay [26]. If the coupling between processes that occur at the microscale and the macroscopic phenomena is tight, then Scheibe et al. [33] advocate the use of hybrid multiscale approaches that combine multiple – and commonly very different – models at different scales. Examples include the use of cloud-resolving models embedded in a global climate model [18], or the simulation of flow through fractured rock [2, 27]. There are mathematical advances in multi-scale modeling [13] that have been effectively applied in material sciences but remain to be fully adapted, developed and implemented in biogeochemical problems. It is also noteworthy that the scale selected to resolve explicitly not only impacts the parameterization, but also the appropriate choice of model formulation, because stochasticity in ecological dynamics observed at the small scale can be averaged out in a coarse grained description [7].

3 Microbially Mediated Biogeochemical Cycling: *The Challenge of Complexity*

The challenge of modeling complexity involves describing the connectivity between dynamical units [4] to reproduce the structural properties of the natural system as well as the interactions between dynamical (sub-)systems, which give rise to emergent system properties. In practice, many terrestrial and marine biogeochemical models contain ad hoc descriptions of biogeochemical networks and the formulation of the processes as represented in the models is far from unique.

One obvious issue is how much detail to include in such a model. To take a specific example, consider the work by Yool [42] who discusses nitrification as part of a plankton ecosystem model that includes phytoplankton (P), zooplankton (Z), detritus (D), ammonium (A) and nitrate (N) as state variables. These compartments are connected via the uptake of nitrate and ammonium by phytoplankton (termed new and regenerated production), grazing of phytoplanktion, predation on the zooplankton and metabolic losses of P and Z and the associated production of detritus, its removal by sinking, ammonium production due the mineralization of detrital material, as well as the conversion of ammonium to nitrate via nitrification. Yool then presents a compilation of eleven distinct rate laws from the literature that describe the kinetics of nitrification, ranging from simple first order substrate dependent rates to formulations that account for the effects of light, temperature, or O_2 availability, with or without thresholds. Notably, these models do not explicitly contain the microbes catalyzing the oxidation of ammonium to nitrate. Thus, Yool [42] presents an expanded model structure, in which nitrifying microbes oxidize ammonium, and then are subject to metabolic losses and grazing by zooplankton. Such a framework allows one to, for example, represent seasonal changes in functional microbial communities. It is easy to imagine further expansion, as nitrification is a multistep process, with nitrite as an extracellular intermediate which in turn also interacts in other nitrogen cycling processes such as denitrification or anammox [5]. The implications of differences in model structures, and of the choices of process descriptions in environmental models are, however, only rarely explored [15, 37]. A rigorous mathematical analysis would be very useful in establishing the validity of such procedures and thus increasing our confidence in the predictive power of model simulations.

One natural approach is to take a network that includes all known relevant processes and make a systematic, mathematical reduction to a simpler system. There is a rich literature on such strategies for dealing with and reducing complexity (e.g., Ratto et al. [30]; Ward et al. [40]). The presence of multiple time scales in the biochemical reactions may afford a natural reduction and still render behavior on the different scales. The resulting perturbation theory is singular and falls generally into two categories: (1) when the fast system is oscillatory and the system is reduced by averaging. This is dealt with in, for instance, Pavliotis and Stuart [29]; and (2) when the fast system is saddle-like near the so-called slow manifold. This is considered by Jones [20] and is called geometric singular perturbation theory. This approach has now been developed to a high level of sophistication, see [10], and a range of time scales beyond one fast and one slow can also be considered. Various further reductions based on multiple time scales have been developed by the engineering community and put on a firm foundation in Kaper and Kaper [21]. Nevertheless, since a systematic reduction of a complex system will not always be evident, it is important to develop measures of the quality of representation by the reduced system. This could include the identification of signatures in timeseries (such as critical slowing down or asymmetric fluctuations) or in spatial patterns (e.g., increase in cross-correlation) of ecological data that points to an approach to critical transitions or tipping points [32]. Furthermore, we expect that such timeseries data, including the observed fluctuations, contain information on the adequate model structure describing the natural system. In addition, the analysis of simplified model descriptions [25] may also help identify conditions that merit close scrutiny in more complex model, as well as through field observations.

4   Merging Data and Models: *The Challenge of Heterogeneous and Sparse Observations*

Biogeochemical models aim to represent complex systems, and require a significant level of abstraction to be tractable. As discussed above, they are not unique in structure, and their parameterization has to account for the simplifying assumptions made. Because of the inability to rigorously establish a bottom-up approach based on first principles, it is important to tightly connect the model with observational data. Such data, however, tend to be sparse, and often of a nature that may not match modeled state variables directly. To illustrate this challenge, we use the Deepwater Horizon oil spill as an example.

The injection of large quantities of hydrocarbons (oil and gas) at a water depth of approx. 1500 m at the Macondo Prospect in the Gulf of Mexico between 22 April and 15 July 2010 constituted an unprecedented perturbation of the Gulf ecosystem, which raised the question of the distribution, fate and impact of both oil and gas. Following the first reports of the existence of subsurface hydrocarbon plumes and the detection of O_2 drawdown in the deep water [16], attempts have been made to quantify the magnitude of these plumes. Of particular interest was (i) to understand how much O_2 is consumed, as it may provide insight into the amount of oil and gas removed in the water column [12], and (ii) to identify the process(es) responsible and their environmental controls (e.g., Crespo-Medina et al. [9]), as these plumes were found well below the zone of photosynthesis or the mixed ocean surface layer, i.e. at a depth where O_2 is only delivered through (slow) vertical mixing. One of the challenges in quantifying the magnitude of deep water O_2 consumption is that direct biogeochemical observations are sparse. Even in the aftermath of the Deepwater Horizon incident, which elicited a dense deployment of research vessels and autonomous vehicles (see e.g., https://data.gulfresearchinitiative.org and http://www.nodc.noaa.gov/deepwaterhorizon/), typical data coverage during a cruise was on the order of one water column profile per $100\,km^2$, with higher resolution along ship tracks. Thus, direct observations of O_2 depletion from shipboard profiles provide limited insight into the spatial variability below scale lengths of kilometers, even when sophisticated interpolation techniques and geostatistics are employed [22, 24].

This patchiness in the data can be constrained by observational tools such as gliders, autonomous underwater vehicles equipped with appropriate sensors recording the physical or chemical composition of the water, or with reaction-transport models that combine flow dynamics and hydrocarbon breakdown, filling in gaps in observations by building on the knowledge of underlying fluid dynamical, chemical and microbiological processes. Such a model has been established by Valentine et al. [39], who simulated hydrocarbon distribution, bacterial metabolism and growth in 2D flow at the depth of the hydrocarbon plume over a period of 5 months following the onset of the spill. The flow model resolution was set to about 4 km, and the biogeochemical model consisted of 26 hydrocarbon compounds decomposed into 26 intermediate chemicals and CO_2 through two microbial populations each, resulting in a state variable for each chemical and microbial population or operational metabolic types (OMT; see supplemental material in Valentine et al. [39]). Biogeochemical parameters were largely set based on estimated microbial activity in the wake of the spill, without optimization. The consumption of O_2 for each reaction step was computed and the resulting O_2 distributions were compared to observations. The results of these simulations revealed substantial microbial dynamics, and a succession of different OMTs, with a strong impact of mixing and flow on microbial hydrocarbon processing. While subsequent work showed that a higher grid resolution would be needed to fully capture the flow dynamics in the plume layer [6], which may affect the computed microbial transformations, Valentine et al. [39] achieved a close match to observed O_2 anomalies.

Data are being used in two different ways in the implementation of these reaction-transport models. Both are fundamental to the final objective but also raise quite distinct challenges.

1. Data are incorporated routinely in the computational model of the fluid state, which gives the underlying velocity field for transport. This renders a state estimation of the fluid independently of the biogeochemistry.
   Key Challenge: the dimensions of such a fluid-dynamical model can be very high, of the order of 10^6 or higher. The methods that can be implemented at this scale involve linearization at some level, for instance Kalman filtering or least-squares approximations in the variational approach. Bayesian methods, such as Markov

Chain Monte-Carlo or particle filtering, work effectively while respecting the underlying nonlinearities. They break down, however, in such high dimensions. There is therefore a fundamental, and unresolved, tension between addressing nonlinearity and dimension that presents an enormous mathematical challenge.

2. Observations of various concentrations and biogeochemical events are used to form the underlying reaction model and to calibrate the reactions and their rates. Since not all microbial players are even known, let alone included in a model, the parameters used in the kinetic description of biogeochemical transformations usually integrate the effect of multiple factors. Moreover, concentrations, commonly used to validate models, reflect the *net* effects of all sources and sinks. To ensure a model captures the underlying processes correctly direct rate measurements and their use in assessing model performance are important.

   Key Challenge: While measured concentrations typically directly correspond to state variables, measurements of process rates that capture the underlying dynamics are rare. Furthermore, other datasets are non-standard in that they represent proxies for processes part of a biogeochemical model. For instance, genomic data on the presence of a given species provides insight into dominant taxonomic groups; the presence of functional genes catalyzing a process of biogeochemical significance indicates the potential for that process to occur; expression data (transcriptomics) or the presence of enzymes (proteomics) in turn points to that process being active. From a mathematical standpoint, this means there is a highly complex observation operator as the relationship between measurements and the state variables is far from direct. While mathematical and statistical techniques can be adapted to such a situation, it is not clear that they will be optimally suited to parameter estimation in these biogeochemical settings.

A critical issue is whether these two manifestations of data assimilation end up being implemented consistently. For instance, it is possible that errors on the transport side are being compensated by adjustments in the biogeochemistry. A unified approach is advocated by Dowd et al. [11], who give a statistical framework for combined data assimilation. Despite its being theoretically sound, their framework will pose many challenges in being implemented, not least of which will be the effective statistical sampling of the high dimensional distributions involved. A hybrid strategy is formulated by [31] that holds promise for problems such as occur in marine biogeochemical models. The many alternatives for model formulations, the heterogeneity of the data available, and the sparsity of the data pose significant challenges for data assimilation. However, also with the establishment of new data collecting infrastructure (e.g. ocean observatories, http://oceanobservatories. org, http://www.emso-eu.org), there is a need to integrate these observations into a comprehensive quantitative framework.

5 Summary and Mathematical Challenges

We have discussed three areas where the input of mathematical analysis can have significant impact on both the tools available to the field scientists and our overall understanding of the underlying physics and biology.

1. Connecting the microbial world to the environment characterization, and ultimately feedbacks of microbial activity to climate conditions requires crossing the gap between the micrometer to kilometer scales. This requires a multiscale analysis and use of scale-appropriate equations. There is a significant body of mathematical work that has been developed within the framework of materials and fluids that could have promising application in the area of biogeochemistry. These include: equation-free methods [23] rigorous homogenization strategies [29] and multi-scale computational techniques [13].

2. The problem of complexity comes down to the fact that many model formulations can reasonably describe the same environmental system. The implication of this, often somewhat arbitrary, choice of model structure is rarely explored. A systematic and objective approach to model reduction is needed. Complex models do not always outperform simpler descriptions [15], but the simpler descriptions need to be known to be faithful to the full dynamics even under broader circumstances than have been previously observed. The separation of time scales among the rates of the various processes may afford a reduction through the techniques of singular perturbation theory [21] as might the functional grouping of species and the use of network theory [28]. There is a wealth of literature on chemical reaction network theory that has been applied in other areas of biology to detect unanticipated simplicity [36], which may be useful here also.

3. Biogeochemical data are sparse, despite the reality of ever increasing numbers and sizes of datasets. Even large environmental metagenomic or transcriptomic datasets can be considered sparse as they typically only provide information on a single location at one point in time, yet expression varies with time and space. Data assimilation techniques that make optimal use of the data available are critical to improve predictive power, but remain underdeveloped in this area. A specific challenge in data assimilation in biogeochemical models is that the model output and the data measured will, in general, not correspond. It requires the integration of semi-quantitative model – data comparisons (for example a comparison of transcriptomic data to process rates), with those of model variables that are measured directly in the field (e.g. metabolite concentration or process rates). Moreover the assimilation of data in the fluid, to determine the state, and in the biogeochemistry, to determine the reaction parameters, needs to be carried out in concert. This is an enormous challenge mathematically as it raises the underlying tension in data assimilation between nonlinearity and dimension as the only effective strategies in high dimensions involve some form of linear approximation.

These areas offer extraordinary opportunities for further mathematical developments, yet to have a strong impact on the quantitative analysis of Earth system behavior, they require a close collaboration between geoscientists and mathematicians. This is a challenging undertaking, in part due to disciplinary culture and language barriers. Training programs that foster collaboration between the next generations of scientists in the two fields may build the basis to advance the field.

Acknowledgements We would like to thank the organizers of the workshop on "Mathematical paradigms of climate change" for their hospitality and the opportunity to discuss some of the biogeochemical aspects in climate-relevant model formulations. This proceeding also benefitted from subsequent discussions at the US-NSF Workshop on 'Expanding the role of reactive transport modeling within the Biogeochemical Sciences', the Gordon Conference on Marine Microbes, and from discussions with A. Bracco and S. Joye. This is ECOGIG contribution #346.

References

1. Arrigo, K.R.: Marine microorganisms and global nutrient cycles. Nature **437**, 349–355 (2005)
2. Battiato, I., Tartakovsky, D.M., Tartakovsky, A.M., Scheibe, T.D.: Hybrid models of reactive transport in porous and fractured media. Adv. Water Resour. **34**, 1140–1150 (2011)
3. Berglund, N., Gentz, B.: Metastability in simple climate models: pathwise analysis of slowly driven Langevin equations. Stoch. Dyn. **2**, 327–356 (2002)
4. Boccaletti, S., Latora, V., Moreno, Y., Chavez, M., Hwang, D.U.: Complex networks: structure and dynamics. Phys. Rep. **424**, 175–308 (2006)
5. Brandes, J.A., Devol, A.H., Deutsch, C.: New developments in the marine nitrogen cycle. Chem. Rev. **107**, 577–589 (2007)
6. Cardona, Y., Bracco, A.: Predictability of mesoscale circulation throughout the water column in the Gulf of Mexico. In: Deep Sea Research Part II: Topical Studies in Oceanography, vol. 129, pp. 332–349. http://dx.doi.org/10.1016/j.dsr2.2014.01.008 (2016)
7. Chave, J.: The problem of pattern and scale in ecology: what have we learned in 20 years? Ecol. Lett. **16**, 4–16 (2013)
8. Ciais, P., Sabine, C., Bala, G., Bopp, L., Brovkin, V., Canadell, J., Chhabra, A., DeFries, R., Galloway, J., Heimann, M., Jones, C., Le Quéré, C., Myneni, R.B., Piao, S., Thornton, P.: Carbon and other biogeochemical cycles. In: Stocker, T.F., Qin, D., Plattner, G.-K., Tignor, M., Allen, S.K., Boschung, J., Nauels, A., Xia, Y., Bex, V., Midgley, P.M. (eds.) Climate Change 2013: The Physical Science Basis. Contribution of Working Group I to the Fifth Assessment Report of the Intergovernmental Panel on Climate Change, pp. 465–570. Cambridge University Press, Cambridge (2013)
9. Crespo-Medina, M., Meile, C.D., Hunter, K.S., Diercks, A.R., Asper, V.L., Orphan, V.J., Tavormina, P.L., Nigro, L.M., Battles, J.J., Chanton, J.P., Shiller, A.M., Joung, D.J., Amon, R.M.W., Bracco, A., Montoya, J.P., Villareal, T.A., Wood, A.M., Joye, S.B.: The rise and fall of methanotrophy following a deepwater oil-well blowout. Nat. Geosci. **7**, 423–427 (2014)
10. Doelman, A., Seawalt, L., Zagaris, A.: The effect of slow processes on emerging spatio-temporal patterns. Chaos **25**, 036408 (2015)
11. Dowd, M., Jones, E., Parslow, J.: A statistical overview and perspectives on data assimilation for marine biogeochemical models. Environmetrics **25**, 203–213 (2014)
12. Du, M., Kessler, J.D.: Assessment of the spatial and temporal variability of bulk hydrocarbon respiration following the deepwater horizon oil spill. Environ. Sci. Technol. **46**, 10499–10507 (2012)
13. Weinan, E.: Principles of Multiscale Modeling. Cambridge University Press, Cambridge (2011)

14. Falkowski, P., Scholes, R.J., Boyle, E., Canadell, J., Canfield, D., Elser, J., Gruber, N., Hibbard, K., Högberg, P., Linder, S., MacKenzie, F.T., Moore III, B., Pedersen, T., Rosenthal, Y., Seitzinger, S., Smetacek, V., Steffen, W.: The global carbon cycle: a test of our knowledge of earth as a system. Science **290**, 291–296 (2000)
15. Friedrichs, M.A.M., Dusenberry, J.A., Anderson, L.A., Armstrong, R.A., Chai, F., Christian, J.R., Doney, S.C., Dunne, J., Fujii, M., Hood, R., McGillicuddy, D.J., Moore, J.K., Schartau, M., Spitz, Y.H., Wiggert, J.D.: Assessment of skill and portability in regional marine biogeochemical models: Role of multiple planktonic groups. J. Geophys. Res.-Oceans **112**, 22 (2007)
16. Gillis, J.: Giant Plumes of Oil Forming Under the Gulf, The New York Times, 16 May 2010
17. Gramling, C.M., Harvey, C.F., Meigs, L.C.: Reactive transport in porous media: a comparison of model prediction with laboratory visualization. Environ. Sci. Technol. **36**, 2508–2514 (2002)
18. Gustafson, W.I., Berg, L.K., Easter, R.C., Ghan, S.J.: The Explicit-Cloud Parameterized-Pollutant hybrid approach for aerosol-cloud interactions in multiscale modeling framework models: tracer transport results. Environ. Res. Lett. **3**, 7 (2008)
19. Hajima, T., Kawamiya, M., Watanabe, M., Kato, E., Tachiiri, K., Sugiyama, M., Watanabe, S., Okajima, H., Ito, A.: Modeling in Earth system science up to and beyond IPCC AR5. Prog. Earth Planet. Sci. **1**, 1–25 (2014)
20. Jones, C.K.R.T.: Geometric singular perturbation theory. In: Johnson, R. (ed.) Dynamical Systems, pp. 44–118. Springer, Berlin (1995)
21. Kaper, T., Kaper, H.: Asymptotic analysis of two reduction methods for chemical kinetics. Physica D **165**, 66–93 (2002)
22. Kessler, J.D., Valentine, D.L., Redmond, M.C., Du, M.R., Chan, E.W., Mendes, S.D., Quiroz, E.W., Villanueva, C.J., Shusta, S.S., Werra, L.M., Yvon-Lewis, S.A., Weber, T.C.: A persistent oxygen anomaly reveals the fate of spilled methane in the deep gulf of Mexico. Science **331**, 312–315 (2011)
23. Kevrekidis, I.G., Gear, C.W., Hummer, G.: Equation-free: the computer-aided analysis of complex multiscale systems. AIChE J. **50**, 1346–1355 (2004)
24. Lai, M.-J., Meile, C.: Scattered data interpolation with nonnegative preservation using bivariate splines and its application. Comput. Aided Geom. Des. **34**, 37–48 (2015)
25. Larsen, L., Thomas, C., Eppinga, M., Coulthard, T.: Exploratory modeling: extracting causality from complexity. Eos, Trans. Am. Geophys. Union **95**, 285–286 (2014)
26. Meile, C., Tuncay, K.: Scale dependence of reaction rates in porous media. Adv. Water Resour. **29**, 62–71 (2006)
27. Neuman, S.P.: Trends, prospects and challenges in quantifying flow and transport through fractured rocks. Hydrogeol. J. **13**, 124–147 (2005)
28. Newman, M.E.J.: Modularity and community structure in networks. Proc. Natl. Acad. Sci. U. S. A. **103**, 8577–8582 (2006)
29. Pavliotis, G.A., Stuart, A.: Multiscale Methods: Averaging and Homogenization. Springer, New York (2008)
30. Ratto, M., Castelletti, A., Pagano, A.: Emulation techniques for the reduction and sensitivity analysis of complex environmental models. Environ. Model. Softw. **34**, 1–4 (2012)
31. Santitissadeekorn, N., Jones, C.: Two-stage filtering for joint state-parameter estimation. Mon. Weather Rev. **143**, 2028–2042 (2015)
32. Scheffer, M., Bascompte, J., Brock, W.A., Brovkin, V., Carpenter, S.R., Dakos, V., Held, H., van Nes, E.H., Rietkerk, M., Sugihara, G.: Early-warning signals for critical transitions. Nature **461**, 53–59 (2009)
33. Scheibe, T.D., Murphy, E.M., Chen, X., Rice, A.K., Carroll, K.C., Palmer, B.J., Tartakovsky, A.M., Battiato, I., Wood, B.D.: An analysis platform for multiscale hydrogeologic modeling with emphasis on hybrid multiscale methods. Groundwater **53**, 38–56 (2014)
34. Schlesinger, W.H., Bernhardt, E.S.: Biogeochemistry, 3rd edn. Academic Press, New York (2013)
35. Schoepp-Cothenet, B., van Lis, R., Atteia, A., Baymann, F., Capowiez, L., Ducluzeau, A.-L., Duval, S., ten Brink, F., Russell, M.J. and Nitschke, W.: On the universal core of bioenergetics. Biochimica et Biophysica Acta (BBA) – Bioenergetics **1827**, 79–93 (2013)

36. Shinar, G., Feinberg, M.: Concordant chemical reaction networks. Math. Biosci. **240**, 92–113 (2005)
37. Tian, R.C.C.: Toward standard parameterizations in marine biological modeling. Ecol. Model. **193**, 363–386 (2006)
38. Todd-Brown, K.E.O., Hopkins, F.M., Kivlin, S.N., Talbot, J.M., Allison, S.D.: A framework for representing microbial decomposition in coupled climate models. Biogeochemistry **109**, 19–33 (2012)
39. Valentine, D.L., Mezic, I., Macesic, S., Crnjaric-Zic, N., Ivic, S., Hogan, P.J., Fonoberov, V.A., Loire, S.: Dynamic autoinoculation and the microbial ecology of a deep water hydrocarbon irruption. Proc. Natl. Acad. Sci. U. S. A. **109**, 20286–20291 (2012)
40. Ward, B.A., Schartau, M., Oschlies, A., Martin, A.P., Follows, M.J., Anderson, T.R.: When is a biogeochemical model too complex? Objective model reduction and selection for North Atlantic time-series sites. Prog. Oceanogr. **116**, 49–65 (2013)
41. Weston, N.B., Joye, S.B.: Temperature-driven decoupling of key phases of organic matter degradation in marine sediments. Proc. Natl. Acad. Sci. U. S. A. **102**, 17036–17040 (2005)
42. Yool, A.: Modeling the role of nitrification in open ocean productivity and the nitrogen cycle. In: Klotz, M.G. (ed.) Methods in Enzymology: Research on Nitrification and Related Processes, vol. 486, Part A, pp. 3–32. Elsevier Academic Press, San Diego (2011)

Turbulence, Horizontal Convection, and the Ocean's Meridional Overturning Circulation

Francesco Paparella

Abstract Kolmogorov's 1941 theory describes a turbulent flow as one featuring a cascade of characteristic scales and obeying the law of finite dissipation. Most flows having the first property also have the second. But this is by no means a necessary implication. Horizontal convection is a type of buoyancy driven flow that does not obey to the law of finite dissipation (thus is not turbulent in the sense of Kolmogorov) but appears to have most other properties of a turbulent flow. This, and a number of other connected results, have profound implications on the viability of horizontal convection as a convincing metaphor for the ocean's meridional overturning circulation: thermal (or thermohaline) forcing alone cannot reproduce most of the essential features of the observed ocean circulation. Recent research shows that only by taking into account the buoyancy forcing and several sources of mechanical mixing in a rotating reference frame a satisfactory depiction of the ocean circulation arises.

Keywords Turbulence • Kolmogorov's 1941 theory • Horizontal convection • Energy dissipation • Meridional overturning circulation

1 On the Defining Properties of Turbulence

Everybody knows what is turbulence, as long as nobody asks for a precise definition. This disappointing state of affairs is made evident by the overabundant distinct notions of turbulence that may be found in the technical literature. For example, it is not uncommon to find the word "turbulence" as a synonym for "deterministic non-periodic flow". The wisest among the proponents of this conceptualization usually add that it applies to "low Reynolds numbers", a specification that cuts out of the picture most naturally occurring turbulent flows (see e.g. [1] for an early authoritative exposition of this view). For many engineering applications the

F. Paparella (✉)
Dipartimento di Matematica e Fisica "Ennio De Giorgi", Università del Salento, Salento, Italy

Division of Sciences, New York University Abu Dhabi, Abu Dhabi, UAE
e-mail: francesco.paparella@unisalento.it; francesco.paparella@nyu.edu

© Springer International Publishing Switzerland 2016
F. Ancona et al. (eds.), *Mathematical Paradigms of Climate Science*, Springer
INdAM Series 15, DOI 10.1007/978-3-319-39092-5_2

defining property of turbulence is its "diffusivity", namely the ability to cause "rapid mixing and increased rates of momentum, heat, and mass transfer" [2]. This view clashes with the observation that some very simple flows, that no one would qualify as "turbulence", have the same property, for example a steady shear [3]. A fundamental flaw of both characterizations is that they overlook the spatial structure of the flow.

A much better description of turbulence is actually rather old, and goes back to the views of Lewis Fry Richardson: "Big whirls have little whirls that feed on their velocity, and little whirls have lesser whirls, and so on to viscosity – in the molecular sense" [4]. Turbulence is thus a flow characterized by a hierarchical *cascade* of scales of motion, coupled together and extending from the largest eddies (whirls) down to the smallest structures that can survive the homogenizing effects of viscosity. Richardson also showed that a cascade leads to enhanced diffusivity, although of a form that cannot be reduced simply to Fick's law, because of the lack of separation between macro and micro scales [5].

Experimentally, in many cases the cascade appears to have a self-similar nature, described by the well-known two-thirds law

$$\left\langle \|u(x, t) - u(y, t)\|^2 \right\rangle \propto \|x - y\|^{2/3} \tag{1}$$

where u is the velocity field of the fluid, the angular brackets denote space and time averages, and the distance $\|x - y\|$ must be smaller than the typical size of the largest eddies and larger than the viscosity cut-off scale. The most striking fact in (1) is that the average velocity fluctuation between nearby points depends only on the distance between them. That is, velocity fluctuations are homogeneous and isotropic.

The best theoretical framework for understanding turbulence was given in 1941 by Andrey Nikolaevich Kolmogorov. Rather than attempting to deduce everything directly from the Navier-Stokes equations, Kolmogorov's theory postulates that there exist flows (that is, solutions of the equations) that satisfy a small set of assumptions. Leveraging on these, several quantitative properties of the flow may then be deduced (e.g. the value 2/3 of the exponent which appears in (1), or the size of the viscosity cut-off scale).

Thus, a precise definition of "turbulent flows" would be: "those (if any) that satisfy Kolmogorov's assumptions". The assumptions are: homogeneity and isotropy at scales smaller than those of the largest eddies, the presence of a self-similar cascade, and the so-called "law of finite dissipation", that, as expressed by Uriel Frisch (see [6] which contains the most readable and complete account of Kolmogorov's 1941 theory in the English language), reads:

> In the limit of infinite Reynolds number, a turbulent flow has a finite, nonvanishing mean rate of [kinetic energy] dissipation ε per unit mass.

An empirical example showing the validity of this law is the fact that the power required to move a solid object at constant speed through a fluid is independent of the viscosity, if the speed is sufficiently large. More precisely, defining the Reynolds number

$$Re = \frac{UL}{\nu} \tag{2}$$

where U is the speed of the object, L is its characteristic size, and ν is the kinematic viscosity of the fluid, the power needed to move the object at constant speed becomes independent of Re when $Re \to \infty$. The object transmits this power to the fluid in its wake, where it stirs a cascade of ever-changing and interacting eddies and whirls.

The law of finite dissipation creates an unforeseen difficulty in our definition of a turbulent flow. In fact, it is always possible, in principle, to ascertain if velocity fluctuations in a flow are homogeneous, isotropic and self-similar. But only by performing multiple experiments with varying Reynolds numbers one may verify whether the law of finite dissipation holds or not. Thus, turbulence would appear to be a property of a *class of flows*, rather than of an individual one.

In many cases of practical interest, if there is a homogeneous, isotropic and self-similar cascade, then, for large enough Reynolds numbers, ε becomes independent of Re. For this reason the law of finite dissipation is often overlooked, and the cascade is taken as the defining property of turbulence. Yet, in Kolmogorov's theory those are independent assumptions, and there is no theoretical argument hinting that one should cause the other. It is therefore interesting to look for flows that appear turbulent (in the sense of having a cascade) but do not obey to the law of finite dissipation.

For mechanically forced flows, the outstanding candidate is the wall shear flow, where the fluid is confined in a half space, has zero velocity on the boundary plane, and tends to a constant velocity, parallel to the plane, as the distance from the plane tends to infinity. For this kind of flow, a long-standing conjecture (which goes back to Prandtl and von Karman) implies that $\varepsilon \propto (\log Re)^{-2}$. Such a weak deviation from constancy is difficult to be detected experimentally, and is also elusive to be proven rigorously [7].

Thermally forced flows offer a better chance of finding a turbulent-looking flow which is not subject to the law of finite dissipation. In the case of Rayleigh-Bénard convection, where the fluid is kept in motion by the temperature difference between two plane parallel plates, it has been hypothesized the existence of an "ultimate regime" in which the heat flux (and consequently the kinetic energy dissipation) becomes independent of the molecular parameters (namely viscosity and thermal diffusivity). In this regime the fluid would be turbulent in the sense of Kolmogorov. Laboratory experiments have had surprising difficulties at attaining this regime, with contradictory results [8, 9]. Mathematical investigations have also faced formidable difficulties, and, so far, it has been impossible either to prove or disprove the conjecture of the existence of an ultimate regime. Thus, the question whether Rayleigh-Bénard convection obeys the law of finite dissipation is still open [10].

Surprisingly, for *horizontal convection*, (a slightly different type of thermally forced flow) it is instead very simple to prove that the law of finite dissipation does not hold. The fact that horizontal convection is the most obvious model for the meridional circulation of the oceans implies that this result sets important constraints on climatological theories.

2   Horizontal Convection

Let us consider a body of fluid confined laterally and from below by perfectly insulating, rigid boundaries and from above by a surface where the potential of the external forces acting on the fluid is constant (for geophysical applications that is the so-called *geopotential,* defined as the sum of the gravitational and centrifugal potentials). On this top surface the *buoyancy b* is prescribed. Buoyancy is a convenient quantity to use in place of the density ρ, and is defined as

$$b = g\left(1 - \frac{\rho}{\rho_o}\right) \tag{3}$$

where g is the acceleration of gravity and ρ_o is a reference density (in the following it will be the maximum density of the initial condition, or of the top boundary condition, whichever is the greatest). A laboratory experimental set-up may look like Fig. 1 where the top surface is a horizontal plane. In this setting the equations of motion read

$$\frac{Du}{Dt} + 2\boldsymbol{\Omega} \times \boldsymbol{u} + \nabla p = b\hat{z} + \nu\nabla^2\boldsymbol{u}, \tag{4a}$$

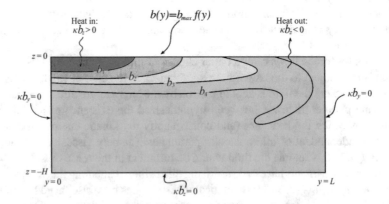

Fig. 1 Horizontal convection set-up. The lateral and the lower boundaries are insulating. On the top surface the temperature is prescribed. Sinking occurs where the top temperature is lower than that of the fluid underneath. After a transient, during which the box fills with cold fluid, sinking happens only around the coldest part of the top boundary (Figure adapted from [13])

$$\frac{Db}{Dt} = \kappa \nabla^2 b, \tag{4b}$$

$$\nabla \cdot \boldsymbol{u} = 0, \tag{4c}$$

where $D/Dt := (\partial/\partial t + \boldsymbol{u} \cdot \nabla)$ is the material derivative; $\boldsymbol{u}$ is the velocity field of the fluid, and (u, v, w) are its components in Cartesian coordinates; $\boldsymbol{\Omega}$ is the angular velocity of the rotating container (or of the Earth, in the case of the oceans); p is the pressure; $\hat{z}$ is a unit vector antiparallel to the gravity acceleration; v is the kinematic viscosity of the fluid; κ is the buoyancy diffusivity of the fluid. Here we are using the well-known Boussinesq approximation that treats the fluid as if it were perfectly incompressible, while allowing buoyancy inhomogeneities to cause motion in the fluid. The buoyancy equation is implicitly assuming that there is only one buoyancy-changing scalar transported by the fluid (e.g., temperature). When two or more buoyancy-changing scalars are present simultaneously, and they have different diffusivity (e.g. temperature and salinity), then an equation for buoyancy of the form above cannot be written. A generalization to the case of multiple scalars will be discussed in the following.

The boundary conditions for the buoyancy are

$$b(\boldsymbol{x}, t)|_{top\ surface} = b_{max} f(x, y) \tag{5a}$$

$$\nabla b(\boldsymbol{x}, t) \cdot \hat{\boldsymbol{n}}|_{bottom,\ sides} = 0 \tag{5b}$$

where $\hat{\boldsymbol{n}}$ is the outward unit vector normal to the boundary, and f is a function bounded between zero and one. Appropriate boundary conditions for the momentum are

$$v \left(\frac{\partial u}{\partial z} \hat{\boldsymbol{x}} + \frac{\partial v}{\partial z} \hat{\boldsymbol{y}} \right) \bigg|_{top\ surface} = \boldsymbol{\tau}_s(x, y, t); \quad w|_{top\ surface} = 0 \tag{6a}$$

$$\boldsymbol{u}(\boldsymbol{x}, t)|_{bottom,\ sides} = 0 \tag{6b}$$

where τ_s is a tangential stress applied to the top surface (e.g. the wind blowing over the ocean). If the last boundary condition is substituted with stress-free boundary conditions, all the results mentioned in the following continue to hold.

As long as the solution of this problem exists and is sufficiently smooth,[1] then the buoyancy is subject to a maximum principle: except, possibly, for an initial transient,

---

[1]Existence and smoothness of the solutions of the incompressible Navier-Stokes equations is the subject of one of the "Millennium Prize Problems" and shall not be further discussed here. In the following it is just assumed that the solution exists, is unique, and is as smooth as required by the manipulations that are going to be performed.

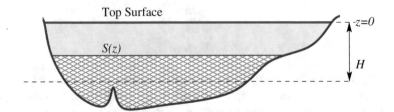

Fig. 2 There is no net flux across the constant-geopotential surface S. The depth H is the ratio between the volume of the fluid and the area of the *top surface*

buoyancy is bounded between the maximum and minimum of the top boundary condition

$$\max_x (b(x, t)) = \max_x \left(b|_{top\ surface}\right).$$
$$\min_x (b(x, t)) = \min_x \left(b|_{top\ surface}\right) \tag{7}$$

The defining characteristic of horizontal convection, with respect to other forms of convection, such as Rayleigh-Bénard's, is that there is no net flux of buoyancy across horizontal surfaces. In fact, let us consider a surface S of constant geopotential, cutting across the fluid. Let us denote with $\langle \cdot \rangle_S$ the operation of averaging in time and over the volume of fluid lying below S (the hatched region in Fig. 2). A straightforward calculation involving the use of the Gauss-Green theorem, shows that the expression

$$\left\langle \frac{Db}{Dt} = \kappa \nabla^2 b \right\rangle_S \tag{8}$$

is equivalent to

$$\overline{wb - \kappa b_z}^S = 0 \tag{9}$$

where the overline denotes an average over time and over the surface S, and b_z is a shorthand for $\partial b/\partial z$. The quantity $wb - \kappa b_z$ is the sum of the advective and diffusive vertical buoyancy fluxes.

The proof that the law of finite dissipation does not hold for horizontal convection follows as a surprising consequence of the energy balance of the fluid. An expression for the kinetic energy is obtained by multiplying Eq. (4a) by u and averaging

$$\left\langle u \cdot \left(\frac{Du}{Dt} + 2\boldsymbol{\Omega} \times u + \nabla p = b\hat{z} + \nu \nabla^2 u \right) \right\rangle. \tag{10}$$

Here $\langle \cdot \rangle$ denotes averaging in time and over the whole domain filled by the fluid. The Coriolis term obviously does not perform mechanical work, and drops out of the average. Integration by parts and the Gauss-Green theorem show that the transport and the pressure terms also average to zero. Only the terms on the right-hand side give a non zero contribution, so that the above expression can be rearranged as

$$\underbrace{\nu \left\langle |\nabla u|^2 + |\nabla v|^2 + |\nabla w|^2 \right\rangle}_{=\varepsilon} = \langle wb \rangle + \frac{\overline{\boldsymbol{u}_s \cdot \boldsymbol{\tau}_s}}{H} \tag{11}$$

where in the last term of the above expression the overline denotes an average over time and over the top surface; $\boldsymbol{u}_s$ is the fluid velocity at the top surface; $\boldsymbol{\tau}_s$ is given by the boundary condition (6a); and H is the ratio between the volume of the fluid and the area of the top surface. The first term in the above expression is the mean rate of kinetic energy dissipation,[2] that is, Kolmogorov's ε.

In the absence of a surface stress ($\boldsymbol{\tau}_s = \boldsymbol{0}$), there is a positive correlation between vertical velocity and buoyancy ($\langle wb \rangle \geq 0$). In other words, lighter fluid must generally move upward, and heavier fluid, for the most part, must go downward. In the absence of boundary conditions that maintain a persistent imbalance in the distribution of buoyancy, this restratification process would move downward the overall center of mass of the system. Therefore, the kinetic energy of the fluid, dissipated by viscosity, has to be created by converting the gravitational potential energy stored in the buoyancy field.

Using the definition (3) of buoyancy, the potential energy of the fluid may be written as

$$E_p = -\rho_o \int_V zb \, d\boldsymbol{x} \tag{12}$$

where V is the volume occupied by the fluid. Because of obvious physical constraints, this is a bounded function of time. Thus, the time average of its rate of change must approach zero in the long run. Therefore, using the buoyancy equation, and after the usual integration by parts, we obtain

$$0 = \lim_{t \to \infty} \left[\frac{1}{t} \int_0^t \frac{d}{ds} \left(\frac{E_p(s)}{\rho_o} \right) ds \right] = -\langle wb \rangle + \kappa \left\langle \frac{\partial b}{\partial z} \right\rangle. \tag{13}$$

A comparison of this expression with Eq. (11) shows that the term $\langle wb \rangle$ quantifies the conversion of potential energy into kinetic energy. Potential energy, in turn, is continuously restored by the boundary conditions. This assert becomes evident by using the Gauss-Green theorem on the last term of (13). Inserting (13) into (11) and

---

[2]It can be shown (see e.g. [11, ch. 4.13][12, ch. 3.4]) that the kinetic energy dissipation rate per unit mass at a point in an incompressible fluid is $\phi = 2\nu e_{ij} e_{ij}$, where $e_{ij} = \left(\partial_{x_j} u_i + \partial_{x_i} u_j \right)$ is the deformation tensor. By averaging ϕ, after an integration by parts, and exploiting incompressibility and the boundary conditions, one obtains the above expression for ε.

using (5a), (5b), for the case of no surface stress we finally have the inequality

$$\varepsilon < \frac{\kappa\, b_{max}}{H} = \frac{\nu\, b_{max}}{Pr\, H} \tag{14}$$

where $Pr = \nu/\kappa$ is the Prandtl number. Therefore, at constant Prandtl number, the energy dissipation rate ε of horizontal convection vanishes for vanishing viscosity. Horizontal convection is non-turbulent in the sense of Kolmogorov [13].

To many people the qualification "non-turbulent flow" conveys the idea of a flow which is either steady or has a very simple time dependence and layered, regular spatial patterns (a "laminar" flow). This certainly was the idea of J. W. Sandström, a Swedish oceanographer who, in the early twentieth century, conducted the first scientific investigations on horizontal convection (including the case with a surface stress). His views may be summarized by the statement

> A circulation can develop from thermal causes only if the level of the heat source lies below the level of the cold source (as quoted in [14]).

This is often called Sandström's "theorem", although it has been long recognized that such a statement, taken at face value, is false, because in the "proof" Sandström did not account for the effects of thermal diffusion [14, 15].

However, Sandström's statement was considered to be essentially right for practical applications [16]. A feeling supported by the earliest numerical simulations of horizontal convection, which showed weak, steady, laminar solutions [17–19]. After an initial transient, the flow settled on a steady pattern characterized by a thermal boundary layer close to the top surface, an interior with low, almost constant buoyancy, and a descending plume connecting the top surface to the interior in correspondence of the lowest buoyancy imposed at the boundary (Fig. 1). The largest speeds were reached along the boundary layer and in the plume, the interior being characterized by a slow, diffused vertical return flow. (For a review on the phenomenology of horizontal convection [20].) The idea that the flow has to organize itself in a very small sinking region and a broadly distributed return flow was already argued by H. Stommel with the use of an ingenious idealized model [21].

More recent calculations showed that unsteady, aperiodic solutions are found [13] if the Rayleigh number

$$Ra = \frac{L^3 b_{max}}{\nu\kappa} \tag{15}$$

is large enough. The boundary layer and the plume are still clearly identifiable; but the interior is not stagnant anymore. Finally, fully three-dimensional, high-resolution numerical calculations showed that unsteady, aperiodic solutions develop a cascade (Fig. 3), as well as other geometrical statistics that are usually associated with genuinely turbulent flows [22]. Therefore, horizontal convection appears to be the first example of a class of flows that exhibit a Richardson-like cascade, but do

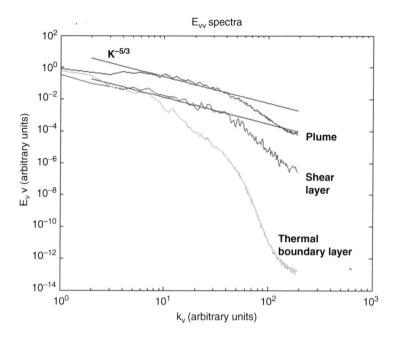

Fig. 3 Power spectra in a high-resolution numerical simulation of unsteady horizontal convection [22]. A $-5/3$ power law scaling is the spectral equivalent of the $2/3$ law (1) [6, cap.5.1] (Figure courtesy of Alberto Scotti)

not obey the law of finite dissipation. Cascade and finite dissipation are independent phenomena.

This independence is further stressed by the fact that the inequality (14) does not imply a bound on the length of the cascade. In fact, in the limit $\nu \to 0$, assuming that the dissipation of kinetic energy occurs homogeneously throughout a large portion of the domain, rather than becoming concentrated only in regions of vanishingly small volume, then the scale η where viscosity arrests the cascade (Kolmogorov's scale) obeys

$$\eta = \left(\frac{\nu^3}{\varepsilon}\right)^{\frac{1}{4}} > \left(\frac{PrH}{b_{max}}\right)^{\frac{1}{4}} \nu^{\frac{1}{2}}. \tag{16}$$

Therefore, (14) does not imply a lower bound on the scale of dissipation, which may go to zero even if dissipation itself goes to zero [23].

These considerations suggest that we might then reserve the word *turbulence* to flows that exhibits the two-thirds law (1), without being concerned whether they also are part of a family of flows that, in the limit of no viscosity, obeys the law of finite dissipation. A result such as (14) should be then called an ε–*theorem*, rather than a *non-turbulence* theorem.

3   A Brief Survey of Other Results on Horizontal Convection

In the last few years, several additional rigorous results on horizontal convection have been found. Some extend and generalize the ε-theorem (14), other consist in bounds on different important physical quantities.

In order to show the relevance for oceanography, the most obvious progress is to extend (14) to the case of two buoyancy-changing scalars, namely temperature and salinity. This was accomplished by J. Nycander [24] with a technique analogous to that of the previous section. He used an equation of state which is quadratic in temperature and includes pressure effects [25, Ch. 1.6]. This captures, albeit in an approximate form, the oceanographically important effects of thermobaricity (increase of the temperature expansion coefficient with pressure) and cabbeling (increase of the temperature expansion coefficient with temperature itself) which are neglected when using a linear approximation of the equation of state. Nycander's result is the following bound

$$\varepsilon < \frac{g\left(\kappa_T \alpha_{max} \Delta\Theta + \kappa_S \beta \Delta S\right)}{H} \tag{17}$$

where κ_T and κ_S are, respectively, the diffusion coefficients of temperature and salinity; α_{max} is the thermal expansion coefficient that one would have at a point where occurred both the maximum temperature and the maximum pressure of the fluid; β is the haline contraction coefficient; $\Delta\Theta$ and ΔS are, respectively, the difference between the maximum and minimum temperature and salinity occurring in the fluid. It is important to stress that Nycander's ε-theorem was obtained in the contest of the *seawater Boussinesq approximation*, a special extension of the ordinary Boussinesq approximation, which allows for a linear dependence of density on an approximate pressure, that depends only on the geopotential height, and which requires to substitute the thermometric temperature T with the *conservative temperature* Θ, a quantity related to the enthalpy of the fluid [26].

By fully disentangling the thermodynamic subtleties of horizontal convection, which, for example, leads to recognize the link between the potential energy (12) and the enthalpy, M. E. McIntyre obtained an ε-theorem valid for a very large class of equations of state, for curved geopotential surfaces, and for equations including the Dufour and Soret cross-diffusive terms. Although McIntyre's result is valid for the seawater Boussinesq equations, there is the concrete possibility that an ε-theorem exists for a generic compressible fluid.

The mean dissipation rate ε of kinetic energy has a counterpart, which is the average dissipation rate of buoyancy variance:

$$\chi = \kappa \left\langle |\nabla b|^2 \right\rangle. \tag{18}$$

Multiplying (4b) by b, after a volume average, integration by parts, and a time average, using the boundary conditions (5a) and (5b), one finds the identity

$$\chi = \frac{\kappa}{H}\overline{b_{max}f(x,y)\frac{\partial b(x,y,0,t)}{\partial z}} \tag{19}$$

showing that χ is proportional to the correlation between the buoyancy forcing $b_{max}f(x,y)$ imposed at the top surface, and the buoyancy flux $\kappa\partial_z b(x,y,0,t)$ leaving or entering the fluid through the top surface. Thus χ quantifies the effectiveness with which a given buoyancy forcing induces a buoyancy flux through the fluid.

Winters and Young [27] have shown, using the same equations and boundary conditions discussed in Sect. 2, that χ is subject to the following bound

$$\chi < 4.57\frac{\kappa^{\frac{1}{3}}b_{max}^{\frac{7}{3}}}{Pr^{\frac{1}{3}}H}. \tag{20}$$

This is a rigorous result, but its physical meaning is best understood by the following heuristic argument, which leads to a similar inequality.

It is perfectly reasonable to assume that the highest speed of the flow will occur in a layer just below to the top surface, where the boundary conditions impose a large-scale, horizontal buoyancy gradient. If U is the characteristic magnitude of the velocity difference across the layer, then the layer thickness may be estimated as $\delta \sim \nu/U$. The mean dissipation rate of kinetic energy will then be $\varepsilon \sim U^3/H$. Using the ε-theorem (14) we are led to conjecture that

$$U \lesssim (\kappa b_{max})^{\frac{1}{3}}; \qquad \delta \gtrsim \frac{Pr\kappa^{\frac{2}{3}}}{b_{max}^{\frac{1}{3}}}. \tag{21}$$

The magnitude of the vertical gradients of buoyancy at the surface should scale as $\partial_z b \sim b_{max}/\delta$. Thus, from (19) and (21) we have

$$\chi \lesssim \frac{\kappa b_{max}^2}{\delta H} \approx \frac{\kappa^{\frac{1}{3}}b_{max}^{\frac{7}{3}}}{PrH}. \tag{22}$$

The crucial point in this line of thought is that the ε-theorem suggests the existence of a lower bound to the thickness δ of the boundary layer, such that, if Pr is kept constant, then δ cannot shrink to zero faster than $\kappa^{\frac{2}{3}}$. Thus, the limited amount of kinetic energy that the ε-theorem allows to inject into the fluid also poses a severe constraint to the rate of dissipation of buoyancy variance.

A crucially important quantity in Rayleigh-Bénard convection is the Nusselt number, defined as the ratio between the heat flux produced by the convection, and the heat flux of the conductive solution. For horizontal convection this definition is meaningless, because there is no net flux across any horizontal surface (see Eq. 9). In place of the Nusselt number it is then useful to use the following dimensionless functional [13, 28]

$$\Phi(b) = \frac{\left\langle |\nabla b|^2 \right\rangle}{\left\langle |\nabla c|^2 \right\rangle} \tag{23}$$

where c is the *conductive solution*, that is the solution of the harmonic equation $\nabla^2 c = 0$, subject to the same boundary conditions (5a) and (5b) of the buoyancy. It is not difficult to show[3] that $\Phi \geq 1$.

Siggers and collaborators proved that the heat flux Φ is bounded by the $1/3$ power of the Rayleigh number

$$\Phi < C_1 Ra^{\frac{1}{3}}. \tag{24}$$

where C_1 is a constant that depends on the top boundary condition of buoyancy. Interestingly, provided that the definition of Φ is changed to $\Phi = \overline{b\partial_z b}|_{top} / \overline{c\partial_z c}|_{top}$, their proof allows for a large class of boundary conditions in place of (5b), including the case of a non-zero buoyancy flux through the bottom (for the ocean, that would take into account the geothermal heating through the ocean floor). For the same class of boundary conditions, they also show that the ε−theorem (14) remains valid. Finally, observing that the boundary layer thickness is $\delta \sim \Phi^{-1}$, the bound (24) implies that the thickness of the top boundary layer scales according to

$$\delta \gtrsim C_2 Ra^{-\frac{1}{3}} \tag{25}$$

for some other constant C_2.

This rigorous bound is less tight than a much older, heuristic bound, first proposed by H. T. Rossby [29]. The argument is set in two dimensions. Horizontally there is only the 'poleward' coordinate y, along which, at the surface, a varying buoyancy is imposed. The flow is described by a streamfunction ψ, and the velocity field is recovered from its derivatives: $\boldsymbol{u} \equiv (v, w) = (-\partial_z \psi, \partial_y \psi)$. From the two-dimensional version of the Eq. (4a), expressed in terms of the streamfunction, Rossby suggested that the main balance in the boundary layer is likely to be $\partial_y b \sim -\nu \partial_{zzzz} \psi$. Then, denoting with V the typical speeds in the boundary layer, we have $\partial_y b \sim b_{max}/L$, $\partial_{zzzz} \psi \sim V/\delta^3$, and

$$V \sim \frac{b_{max} \delta^3}{\nu L}. \tag{26}$$

---

[3]Integrating by parts and applying the boundary conditions one readily obtains $\langle \nabla b \cdot \nabla c \rangle = \langle \nabla c \cdot \nabla c \rangle$. From $0 \leq \left\langle |\nabla b - \nabla c|^2 \right\rangle$, expanding the square, follows $\Phi \geq 1$.

Analogously, from (4b), Rossby suggested the balance $-\partial_z \psi \partial_y b \sim \kappa \partial_{zz} b$, which leads to

$$V \sim \frac{\kappa L}{\delta^2} \tag{27}$$

Combining these relationships one finds

$$\delta \sim L Ra^{-\frac{1}{5}}, \quad V \sim \frac{\kappa}{L} Ra^{\frac{2}{5}}, \quad \psi_{max} = \kappa Ra^{\frac{1}{5}} \tag{28}$$

where the last one comes from $V \sim \psi_{max}/\delta$, and the maximum ψ_{max} of the streamfunction gives a measure of the strength of the overturning circulation.

The scalings (28) are in excellent agreement with the results of the currently available numerical simulations. This opens the question whether the exponent in the bound (24) is not the optimal one, or there is a transition from Rossby's scaling (28) to the "ultimate" regime (24) around a yet–unattained value of the Rayleigh number. In fact, on the basis of careful, very high Rayleigh number laboratory experiments, Hughes and coworkers have developed an idealized model that departs from the above scaling laws and appears to be in agreement with the experiments when the descending plume is not laminar, and produces a significant entrainment into the fluid at every height [30].

4 Implications for Oceanography and Climate

For pedagogical purposes, the ocean circulation is usually separated in two parts: the so-called wind-driven circulation, that occurs only in the first few hundred meters below the surface, and the meridional overturning circulation, that involves the whole depth of the ocean. The latter is sometimes called "thermohaline" circulation, because it is implicitly assumed that buoyancy differences (due to an uneven distribution of temperature and salinity) completely account for the observed overturning motions. In this view, thermohaline circulation is just a natural occurrence of horizontal convection.

Obviously, the pedagogically convenient view would immediately crumble if Sandström's "theorem" were true. Although this is not the case, because horizontal convection can produce a turbulent cascade in the bulk of the fluid, the bounds discussed above show that Sandström's arguments appears to have some merit, if understood in the following sense: in the limit of vanishing diffusivity, an uneven buoyancy forcing applied along a geopotential surface becomes unable to inject mechanical energy and buoyancy variance into the fluid.

Remarkably, the rate of dissipation of kinetic energy in the ocean is accessible to measurement [31, ch. 6]. A reasonable estimate of the mean rate of kinetic energy dissipation for the whole ocean is $\varepsilon \simeq 10^{-9}\,\mathrm{W\,Kg^{-1}}$, but the inequality (14) yields[4] $\varepsilon \lesssim 10^{-12}\,\mathrm{W\,Kg^{-1}}$. Note that $10^{-9}\,\mathrm{W\,Kg^{-1}}$ is a very small power density, anyhow. It corresponds to the power of a typical hairdryer for each cubic kilometer of ocean water. Thus, the inequality (14) implies that horizontal convection alone would only be able to produce in the ocean truly diminutive power densities. A similar mismatch between the bound and the observations occurs for the rate of dissipation of the buoyancy variance χ [27]. On the other hand, the bound (24) is compatible with the observed poleward heat flux of $2 \cdot 10^{15}\,\mathrm{W}$ [28], but the corresponding estimate (25) for the boundary layer yields unbelievably small values (less than a millimeter). Even Rossby's scalings (28) give unrealistically small results for the thickness of the boundary layer (about 3 m) [32], at the cost of a corresponding drop in the strength of the poleward heat flux.

A puzzling mismatch between the observations and the scenario in which the surface buoyancy inhomogeneities are the main cause of the overturning circulation emerged as early as 1966. At that time W. Munk [33] remarked that if the overturning is characterized by fast sinking localized in small regions and a roughly uniform, widespread upwelling in the rest of the ocean, then, in these places, the vertical profiles of buoyancy would be determined by the balance between the upward vertical transport and the mixing due to eddying motion. Namely, the balance would be

$$w\frac{\partial b}{\partial z} = \kappa\frac{\partial^2 b}{\partial z^2}. \tag{29}$$

where the diffusion coefficient κ must be understood as an *eddy diffusivity*: the effect of the small-scale transport is summarized, using a mixing-length argument, by an enhanced diffusion (see e.g. [11, ch. 13.12]). This yields exponential profiles that are an acceptable fit to the observations, provided that the vertical length scale κ/w is set to values of the order of 1 km. The available knowledge of the volume of water sinking into the abyss, and of the extent of the sinking regions, allows for a rough estimate of the upward return velocity: $w \simeq 10^{-7}\,\mathrm{ms^{-1}}$. This, in turn, implies an eddy diffusivity of $\kappa \simeq 10^{-4}\,\mathrm{m^2 s^1}$, a value so large that it was not believed to be self-generated by the overturning motion, but had to be traced back to some other cause injecting energy into the ocean. Munk suggested that the energy sources would be winds and tides, creating waves propagating into the deep ocean, that would then break and produce the eddying motions whose mixing effect would balance the upwelling. Interestingly, he did not shy away from proposing more exotic mixing mechanisms, suggesting that plankton and nekton could be able to mix the water column while swimming, an idea that has recently been shown to be much less far-fetched than most of us would be ready to believe [34, 35]. A later analysis, although acknowledging that the upwelling is not as uniform as it was thought to be in the 1960s, confirms the overall scenario, and shows that the predicted eddy

---

[4]Using $b_{max} = 5 \cdot 10^{-2}\,\mathrm{m\,s^{-2}}$; $H = 5000\,\mathrm{m}$; $\kappa = 10^{-7}\,\mathrm{m^2 s^{-1}}$.

diffusivities are compatible with the observed average kinetic energy dissipation rate $\varepsilon \simeq 10^{-9} \, \text{W} \, \text{Kg}^{-1}$ [36].

Thus, it appears that the overturning circulation is not determined by the *"pushing down"* of sinking plumes at high latitudes, but by the *"pulling up"* of small scale turbulent mixing due to causes unrelated to the distribution of buoyancy. The pulling-up hypothesis has faced some difficulties, because the observed eddy diffusivity in most of the ocean is about $\kappa \simeq 10^{-5} \, \text{m}^2\text{s}^1$, that is, about one order of magnitude less than what would be required (this was called the *"missing mixing"* problem). However, as is becoming apparent in the last few years, this does not negate that there is a pulling up mechanism, but rather that it is uniformly distributed, rather than geographically localized.

The issue of horizontal convection as a model for the meridional overturning circulation of the oceans has sparked a considerable debate in the literature, mostly focused on the thermodynamic pathways leading from the observed surface buoyancy distribution to the observed dissipation rate of kinetic energy [27, 37–40]. It turns out that horizontal convection alone might justify a vigorous heat transport [41], but this result is of limited oceanographic significance, because a sizable fraction of the poleward heat transport of the oceans seems to occur in the first few hundred meters of depth, where the circulation is directly caused by the wind [42, 43]. In summary, it is essentially accepted that the causes of the overturning circulation cannot be exclusively ascribed to the latitudinal differences in the surface distribution of buoyancy [44, 45]. Clarifying which buoyancy-independent factors contribute to the overturning circulation, and in what amount, is a problem on which important recent progresses have been made, but can't yet be considered fully solved. The best estimates set at about three terawatts the power required by mixing processes for closing the overturning circulation [46].

If a surface stress, or an interior body force close to the surface is added to a horizontal convection set-up, numerical simulations and laboratory experiments show an intensification of the overturning circulation [47–50], but the flow patterns show little resemblance to the oceanic ones. In particular, Rossby-like scalings (and the associated small boundary layers) persist even in the presence of strong surface stresses [32]. An important difference occurs when rotation of the reference frame is taken into account. When $\mathbf{\Omega} \neq 0$ in the momentum equation (4a) the descending plume may undergo *baroclinic instability* [25, ch. 6], a form of instability which is possible only in rotating fluids. This creates a vigorous eddy field that mixes the interior of the domain, and produces a stratified abyss, much more effectively than non-rotating turbulent entrainment could do [51].

Dramatic changes in the meridional circulation patterns emerge only when the combined action of surface stress and rotation is taken into account. The winds unrelentingly blowing from the west to the east all around the continent of Antarctica (the *roaring forties* and the *furious fifties*, as they are known to sailors), because of the Coriolis force, cause a large northward transport of surface waters, that, in turn, causes an uplift of denser waters from depths as large as 2000 m. This process does not require deep mixing processes, and the uplift may happen adiabatically. The net result is that wind energy is converted into potential energy by raising dense water. The potential energy is then converted into vortical motion through baroclininc instability. This mechanism produces a pole-to-pole recirculation and is supported by theory, observations and numerical simulations [52–54]. It accounts for a very large part (possibly all) of the *missing mixing* problem in the upper half of the ocean.

The bottom half of the ocean can't be uplifted by the winds, even those as strong as the furious fifties. The breaking of internal tidal waves in correspondence of rough bottom topography is considered to be a mayor source of mixing, supplying about one terawatt of energy to the ocean [44, 45, 55]. A large amount of kinetic energy is stored in water masses moving in a state of *"geostrophic equilibrium"* (that is, in which pressure is balanced by the Coriolis force). This equilibrium cannot be maintained forever, and is usually lost when interacting with topography, resulting in the radiation of internal waves, which then break and cause mixing. Estimating how much energy is dissipated through loss of geostrophic balance has proven to be very elusive. However, it is likely to be a non-negligible amount on a global scale [56, 57].

Finally, although the role of the meridional overturning circulation as the *"conveyor belt"* that carries heat poleward is being found less important than previously thought (as mentioned before), the deep overturing circulation plays a crucial climatic role as a regulating mechanism of the atmospheric CO_2 concentration. Changes in the meridional overturning circulation, regulated by delicate feedbacks between the extent of the Antarctic sea ice cover, and the concentration of atmospheric CO_2, may be the main oscillator that synchronizes with the astronomical cycles, giving rise to the glacial and interglacial phases [58, 59].

References

1. Ruelle, D.: The Lorenz attractor and the problem of turbulence. In: Turbulence and Navier Stokes Equations. Lecture Notes in Mathematics, Springer, Berlin, vol. 565, pp. 146–158 (1976)
2. Tennekes, H., Lumley, J.L.: A First Course in Turbulence. MIT Press, Cambridge (1972)
3. Young, W.R., Jones, S.: Shear dispersion. Phys. Fluids A **3**, 1087–1101 (1991)
4. Richardson, L.F.: Weather Prediction by Numerical Process. Cambridge University Press, Cambridge (2007). Reprint of the first edition (1922)
5. Richardson, L.F.: Atmospheric diffusion shown on a distance-neighbour graph. Proc. R. Soc. Lond. Ser. A **110**, 709–737 (1926)
6. Frisch, U.: Turbulence: The legacy of A.N. Kolmogorov. Cambridge University Press, Cambridge (1995)
7. Doering, C.R., Spiegel, E.A., Worthing, R.A.: Energy dissipation in a shear layer with suction. Phys. Fluids **12**, 1955–1968 (2000)
8. Niemela, J.J., Skrbek, L., Sreenivasan, K.R., Donnelly, R.J.: Turbulent convection at very high Rayleigh numbers. Nature **404**, 837–840 (2000)
9. He, X., Funfschilling, D., Nobach, H., Bodenschatz, E., Ahlers, G.: Transition to the ultimate state of turbulent Rayleigh-Be´nard convection. Phys. Rev. Lett. **108**, 024502 (2012)
10. Ahlers, G., Grossmann, S., Lohse, D.: Heat transfer and large scale dynamics in turbulent Rayleigh-Bénard convection. Rev. Mod. Phys. **81**, 503–537 (2009)

11. Kundu, P.K., Cohen, I.M.: Fluid Mechanics, 2nd edn. Academic Press, San Diego (2002)
12. Batchelor, G.K.: An Introduction to Fluid Dynamics. First Cambridge Mathematical Library edition. Cambridge University Press, Cambridge (2000)
13. Paparella, F., Young, W.R.: Horizontal convection is non-turbulent. J. Fluid Mech. **466**, 205–214 (2002)
14. Kuhlbrodt, T.: On Sandström's inferences from his tank experiments: a hundred years later. Tellus **60A**, 819–836 (2008)
15. Jeffreys, H.: On fluid motions produced by differences of temperature and humidity. Q. J. R. Meteorol. Soc. **51**, 347–356 (1925)
16. Wunsch, C.: Moon, tides and climate. Nature **405**, 743–744 (2000)
17. Sommerville, R.C.: A non-linear spectral model of convection in a fluid unevenly heated from below. J. Atmos. Sci. **24**, 665–676 (1967)
18. Beardsley, R.C., Festa, J.F.: A numerical model of convection driven by a surface stress and non-uniform horizontal heating. J. Phys. Oceanogr. **2**, 444–455 (1972)
19. Rossby, H.T.: Numerical experiments with a fluid heated non-uniformly from below. Tellus **50A**, 242–257 (1998)
20. Hughes, G.O., Griffiths, R.W.: Horizontal convection. Annu. Rev. Fluid Mech. **40**, 185–208 (2008)
21. Stommel, H.: On the smallness of the sinking regions in the ocean. Proc. Natl. Acad. Sci. **48**, 766–772 (1962)
22. Scotti, A., White, B.: Is horizontal convection really "non-turbulent?". Geophys. Res. Lett. **38**, L21609 (2011)
23. McIntyre, M.E.: On spontaneous imbalance and ocean turbulence: generalizations of the Paparella–Young epsilon theorem. In: Dritschel, D.G. (ed.) Turbulence in the Atmosphere and Oceans. Proceedings of International IUTAM/Newton Institute Workshop held 8–12 Dec 2008, Montreal, pp. 3–15. Springer (2010)
24. Nycander, J.: Horizontal convection with a non-linear equation of state: generalization of a theorem of Paparella and Young. Tellus **62A**, 134–137 (2010)
25. Vallis, G.K.: Atmospheric and Oceanic Fluid Dynamics: Fundamentals and Large-scale Circulation. Cambridge University Press, Cambridge (2006)
26. Young, W.R.: Dynamic Enthalpy, Conservative Temperature, and the Seawater Boussinesq Approximation. J. Phys. Oceanogr. **40**, 394–400 (2010)
27. Winters, K.B., Young, W.R.: Available potential energy and buoyancy variance in horizontal convection. J. Fluid Mech. **629**, 221–230 (2009)
28. Siggers, J.H., Kerswell, R.R., Balmforth, N.J.: Bounds on horizontal convection. J. Fluid Mech. **517**, 55–70 (2004)
29. Rossby, H.T.: On thermal convection driven by non-uniform heating from below: an experimental study. Deep Sea Res. Oceanogr. Abst. **12**, 9–16 (1965)
30. Hughes, G.O., Griffiths, R.W., Mullarney, J.C., Peterson, W.H.: A theoretical model for horizontal convection at high Rayleigh number. J. Fluid Mech. **581**, 251–276 (2007)
31. Thorpe, S.A.: The Turbulent Ocean. Cambridge University Press, Cambridge (2005)
32. Hazewinkel, J., Paparella, F., Young, W.R.: Stressed horizontal convection. J. Fluid Mech. **692**, 317–331 (2012)
33. Munk, W.H.: Abyssal recipes. Deep-Sea Res. **13**, 707–730 (1966)
34. Dewar, W.K., Bingham, R.J., Iverson, R.L., Nowacek, D.P., St. Laurent, L.C., Wiebe, P.H.: Does the marine biosphere mix the ocean? J. Mar. Res. **64**, 541–561 (2006)
35. Dewar, W.K.: A fishy mix. Nature **460**, 581–582 (2009)
36. Munk, W.H., Wunsch, C.: Abyssal recipes II: energetics of tidal and wind mixing. Deep-Sea Res. I **45**, 1977–2010 (1998)
37. Gnanadesikan, A., Slater, R.D., Swathi, P.S., Vallis, G.K.: The energetics of ocean heat transport. J. Clim. **18**, 2604–2616 (2005)
38. Tailleux, R.: On the energetics of stratified turbulent mixing, irreversible thermodynamics, Boussinesq models and the ocean heat engine controversy. J. Fluid Mech. **638**, 339–382 (2009)

39. Hughes, G.O., McC Hogg, A., Griffiths, R.W.: Available potential energy and irreversible mixing in the meridional overturning circulation. J. Phys. Oceanogr. **39**, 3130–3146 (2009)
40. McC Hogg, A., Dijkstra, H.A., Saenz, J.A.: The energetics of a collapsing meridional overturning circulation. J. Phys. Oceanogr. **43**, 1512–1524 (2013)
41. Gayen, B., Griffiths, R.W., Hughes, G.O., Saenz, J.A.: Energetics of horizontal convection. J. Fluid Mech. **716**, R10-1–R10-11 (2013)
42. Boccaletti, G., Ferrari, R., Adcroft, A., Ferreira, D., Marshall, J.: The vertical structure of ocean heat transport. Geophys. Res. Lett. **32**, L10603-1–L10603-4 (2005)
43. Ferrari, R., Ferreira, D.: What processes drive the ocean heat transport? Ocean Model. **38**, 171–186 (2011)
44. Wunsch, C., Ferrari, R.: Vertical mixing, energy, and the general circulation of the oceans. Annu. Rev. Fluid Mech. **36**, 281–314 (2004)
45. Ferrari, R., Wunsch, C.: Ocean circulation kinetic energy: reservoirs, sources, and sinks. Annu. Rev. Fluid Mech. **41**, 253–282 (2009)
46. St Laurent, L., Simmons, H.: Estimates of power consumed by mixing in the ocean interior. J. Clim. **19**, 4877–4890 (2006)
47. Tailleux, R., Rouleau, L.: The effect of mechanical stirring on horizontal convection. Tellus **62A**, 138–153 (2010)
48. Ilıcak, M., Vallis, G.K.: Simulations and scaling of horizontal convection. Tellus **64A**, 18377-1–18377-17 (2012)
49. Whitehead, J.A., Wang, W.: A laboratory model of vertical ocean circulation driven by mixing. J. Phys. Oceanogr. **38**, 1091–1106 (2008)
50. Stewart, K.D., Hughes, G.O., Griffiths, R.W.: The role of turbulent mixing in an overturning circulation maintained by surface buoyancy forcing. J. Phys. Oceanogr. **42**, 1907–1922 (2012)
51. Barkan, R., Winters, K.B., Llewellyn Smith, S.G.: Rotating horizontal convection. J. Fluid Mech. **723**, 556–586 (2013)
52. Marshall, J., Speer, K.: Closure of the meridional overturning circulation through Southern Ocean upwelling. Nat. Geosci. **5**, 171–180 (2012)
53. Wolfe, C.L., Cessi, P.: The adiabatic pole-to-pole overturning circulation. J. Phys. Ocean. **41**, 1795–1810 (2011)
54. Haertel, P., Fedorov, A.: The ventilated ocean. J. Phys. Ocean. **42**, 141–164 (2012)
55. Ferrari, R.: What goes down must come up. Nature **513**, 179–180 (2014)
56. Dewar, W.K., McC Hogg, A.: Topographic inviscid dissipation of balanced flow. Ocean Model. **32**, 1–13 (2010)
57. Nikurashin, M., Ferrari, R.: Global energy conversion rate from geostrophic flows into internal lee waves in the deep ocean. Geophys. Res. Lett. **38**, L08610-1–L08610-6 (2011)
58. Paillard, D., Parrenin, F.: The Antarctic ice sheet and the triggering of deglaciations. Earth Planet. Sci. Lett. **227**, 263–271 (2004)
59. Ferrari, R., Jansen, M.F., Adkins, J.F., Burke, A., Stewart, A.L., Thompson, A.F.: Antarctic sea ice control on ocean circulation in present and glacial climates. PNAS **111**, 8753–8758 (2014)

Part II
Mathematical Techniques

Source Reconstruction by Partial Measurements for a Class of Hyperbolic Systems in Cascade

Fatiha Alabau-Boussouira, Piermarco Cannarsa, and Masahiro Yamamoto

Abstract We consider a system of two inhomogeneous wave equations coupled in cascade. The source terms are of the form $\sigma_1(t)f(x)$, and $\sigma_2(t)g(x)$, where the σ_i's are known functions whereas the sources f and g are unknown and have to be reconstructed. We investigate the reconstruction of these two space-dependent sources from a single boundary measurement of the second component of the state-vector. We prove identification and stability estimates for all sufficiently large times T under a smallness condition on the norm of $(\sigma_1 - \sigma_2)'$ in $L^2([0, T])$ in the class of coupling coefficients that keep a constant sign in the spatial domain. We give sharper conditions if one of the two kernels σ_i's is positive definite. Furthermore, we give examples of coupling coefficients that change sign within the domain for which identification fails. Our approach is based on suitable observability estimates for the corresponding free coupled system established in Alabau-Boussouira (Math Control Signals Syst 26:1–46, 2014; Math Control Relat Fields 5:1–30, 2015) and the approach based on control theory developed in Puel and Yamamoto (Inverse Probl 12:995–1002, 1996).

Keywords Inverse problems • Hyperbolic systems • Source identification • Indirect observability

F. Alabau-Boussouira (✉)
Institut Elie Cartan de Lorraine, UMR-CNRS 7502, Université de Lorraine, Ile du Saulcy, 57045 Metz Cedex 1, France
e-mail: fatiha.alabau@univ-lorraine.fr

P. Cannarsa
Dipartimento di Matematica, Università di Roma Tor Vergata, Via della Ricerca Scientifica 1, 00133 Roma, Italy
e-mail: cannarsa@mat.uniroma2.it

M. Yamamoto
Department of Mathematical Sciences, The University of Tokyo, Komaba, Meguro, Tokyo 153-8914, Japan
e-mail: myama@ms.u-tokyo.ac.jp

© Springer International Publishing Switzerland 2016
F. Ancona et al. (eds.), *Mathematical Paradigms of Climate Science*, Springer
INdAM Series 15, DOI 10.1007/978-3-319-39092-5_3

1 Introduction

The control of reaction-diffusion systems, of coupled hyperbolic systems, or of more complex systems involved in medical, biological, chemical or mechanical applications, or in the climate evolution, is a challenging issue. In environmental engineering, it is important to model vibration phenomena as well as diffusion, and to model coupling effects in the involved structures. Moreover, due to further constraints of cost or practical realizations, we often cannot take data of all the components of the engineering structure because some components are not accessible e.g., in the case where they are underground. We present here a primitive model of such engineering structures, namely system (1), which is a prototype of the longitudinal vibrations of two connected beams. The goal is to determine spatial distribution of external forces only by data of one component, which is requested to evaluate the strength of the structure. Our approach is to give methodological approaches which show that is it possible to reconstruct initial data from boundary measurements on a single unknown.

Let Ω be a bounded open set in $\mathbb{R}^d$ with a sufficiently smooth boundary Γ. We consider the following coupled cascade system

$$
\begin{cases}
\partial_t^2 y_1 - \Delta y_1 = \sigma_1(t)f(x), & t \in (0,T), x \in \Omega, \\
\partial_t^2 y_2 - \Delta y_2 + c(x)y_1 = \sigma_2(t)g(x), & t \in (0,T), x \in \Omega, \\
y_1 = y_2 = 0, & t \in (0,T), x \in \Gamma, \\
y_i(0,\cdot) = \partial_t y_i(0,\cdot) = 0, & \text{in } \Omega, i = 1,2,
\end{cases}
\tag{1}
$$

where $c : \Omega \mapsto \mathbb{R}$ is a given coupling coefficient, $\sigma_i : [0,T] \mapsto \mathbb{R}$ are given known functions for $i = 1,2$, $f, g : \Omega \mapsto \mathbb{R}$ are unknown sources. Our purpose is to reconstruct f and g:

- only through either boundary or locally distributed measurements of y_2;
- under certain hypotheses on σ_1 and σ_2;
- under general geometric situations, in particular when the support of the coupling coefficient does not meet the support of the measurement.

Let us observe that $y_1 \equiv 0$ when $f \equiv 0$, and the reconstruction problem reduces to the usual scalar source reconstruction problem which has been solved under general hypotheses (see e.g. [12] and [11]). Moreover, when $c \equiv 0$, the two above equations are decoupled, so that there is no hope to reconstruct f through a measurement of y_2 only. This indicates that a positive answer to the problem we consider will require some hypotheses on c.

The scope of this paper is to discuss some first results that can be obtained for the above problem as well as interesting open questions that raise naturally, especially when the kernels σ_1 and σ_2 are different. We will give further results for this problem in a forthcoming paper.

We now proceed to describe the key points of our approach and summarize our main results.

1. We relate our source reconstruction problem for the coupled system to a suitable observability problem for an inhomogeneous system. Note that this approach is in the spirit of a former work by Yamamoto [12] for the scalar wave equation.
2. We use the results of the first author [2–5] which give a necessary and sufficient condition for observability using only the trace of the normal derivative of the second component on some part of the boundary. Such a condition holds in the class of essentially bounded coupling coefficients c with constant sign within the spatial domain. The necessary and sufficient condition is formulated in terms of the scalar free wave equation in the usual energy space $H_0^1(\Omega) \times L^2(\Omega)$. Indeed, one can recover the initial data of both components of the free coupled cascade system, if and only if:

   • the considered *boundary observation operator* satisfies the usual observability estimate for the solution of the scalar free wave equation (condition $(A1)$ below), and
   • the *coupling operator*, viewed as an observation operator, satisfies the usual observability estimate for the solution of the scalar free wave equation (condition $(A2)$ below).

3. We give a sufficient condition on the difference $||(\sigma_1 - \sigma_2)'||_{L^2(0,T)}$ to ensure that we can deduce from the observability estimates for the free coupled system, the desired stability estimate for our original inhomogeneous system.
4. Using the negative observability results of [5], we give examples where identification fails when the coupling coefficient changes sign even though $\sigma_1 \equiv \sigma_2$.
5. An important feature, compared to the case of the source reconstruction for a scalar equation, is that we need a smallness condition on $||(\sigma_1 - \sigma_2)'||_{L^2(0,T)}$. We discuss and relax this condition when one of the σ_i's is a positive definite kernel.

We conclude by describing the outline of this paper. In Sect. 2, we discuss the assumptions we make on the data, especially the geometric conditions on the domain and coupling coefficient. Then, we recall the observability results for homogeneous systems that we will need for the proof of our main results. Section 3 is devoted to the well-posedness of system (1). In Sect. 4, we derive our stability estimates for the inverse source problem for system (1). Finally, we provide counterexamples to uniqueness when the sign of the coupling coefficient is not constant in Sect. 5.

2 Preliminaries and Set-Up

2.1 Notation

For $T > 0$ given, we set $Q_T = (0, T) \times \Omega$ and $\Sigma_T = (0, T) \times \Gamma$. We denote by $H_1 := H_0^1(\Omega)$ the usual Sobolev space and $H_{-1} := H^{-1}(\Omega)$ its dual space with respect to the pivot space $L^2(\Omega)$. We also denote by $H_{-2} := H^{-2}(\Omega)$ the dual space of $H^2(\Omega) \cap H_0^1(\Omega)$ with respect to the pivot space $H := L^2(\Omega)$. We respectively denote by $|\cdot|_{0,\Omega}$ the norm on $L^2(\Omega)$, by $|\cdot|_{1,\Omega}$ the H^1-semi-norm (which is a norm on $H_0^1(\Omega)$), and by $|\cdot|_{-1,\Omega}$ the norm on $H^{-1}(\Omega)$. Moreover for $x \in \Gamma$, we denote by $v(x)$ the outward unit normal to Γ at x. We denote by $\mathscr{H}^{d-1}$ the $(d-1)$-dimensional Hausdorff measure in $\mathbb{R}^d$. Moreover we shall denote by $\mathscr{H}_b^{d-1}$, the measure $b(\cdot)\mathscr{H}^{d-1}$ where $b \in L^\infty(\Gamma)$, $b \geq 0$ is a given function, with a support that we shall denote by Γ_1. $L_b^2(\Gamma)$ denotes the set of measurable functions on Γ which are square integrable with respect to the measure $\mathscr{H}_b^{d-1}$.

For vector-valued functions $t \mapsto V(t) = (v_1(t), v_2(t)) \in H_0^1(\Omega) \times L^2(\Omega)$, we define $e_1(V)(t)$ as

$$e_1(V)(t) = \frac{1}{2}\left(|v_1(t)|_{1,\Omega}^2 + |v_2(t)|_{0,\Omega}^2\right),\tag{2}$$

whereas for vector-valued functions $t \mapsto V(t) = (v_1(t), v_2(t)) \in L^2(\Omega) \times H^{-1}(\Omega)$, we define $e_0(V)(t)$ as

$$e_0(V)(t) = \frac{1}{2}\left(|v_1(t)|_{0,\Omega}^2 + |v_2(t)|_{-1,\Omega}^2\right).\tag{3}$$

2.2 Hidden Regularity and Observability Assumptions for the Scalar Wave Equation

Let us consider the scalar wave equation

$$\begin{cases} \partial_t^2 w - \Delta w = 0 \text{ in } Q_T, \\ w = 0 \text{ on } \Sigma_T, \\ (w, \partial_t w)(0) = (w^0, w^1) \text{ in } \Omega, \end{cases}\tag{4}$$

For every $(w^0, w^1) \in H_1 \times H$, there exists a unique solution (in the sense of the semigroup theory) $w \in \mathscr{C}([0,T]; H_1) \cap \mathscr{C}^1([0,T]; H)$, and the solution is smoother for smoother initial data. We recall the following so-called hidden regularity property: for every $T > 0$, there exists $\eta_1(T) > 0$ such that for all $(w^0, w^1) \in H_1 \times H$ the solution w of (4) satisfies the following direct inequality

$$\int_0^T \int_\Gamma \left|\frac{\partial w}{\partial v}\right|^2 d\mathscr{H}_b^{d-1} dt \leq \eta_1(T)e_1(W)(0),\tag{5}$$

where $W = (w, \partial_t w)$. The reverse inequality is called an observability inequality, it holds only for sufficiently large time T and under geometric conditions which may vary depending on the mathematical tools that are used to derive such an inequality (microlocal analysis, multiplier method ...). For the sake of generality, we shall assume that suitable observability assumptions hold for the scalar wave equation and give in the next remarks, examples of sufficient conditions under which they hold. We consider the following assumption on b

$$(A_1) \begin{cases} b \in L^\infty(\Gamma) , b \geqslant 0 \text{ on } \Gamma \text{ is such that }, \\ \exists\, T_0 > 0, \forall\, T > T_0 , \exists\, \gamma_1(T) > 0 \text{ such that} \\ \forall\, (w^0, w^1) \in H_1 \times H , \text{ the solution } w \text{ of (4) satisfies} \\ \int_0^T \int_\Gamma \left| \dfrac{\partial u}{\partial \nu} \right|^2 d\mathcal{H}_b^{d-1}\, dt \geqslant \gamma_1(T) e_1(W)(0) , \end{cases} \qquad (6)$$

and the following assumption on c

$$(A2) \begin{cases} c \in L^\infty(\Omega) , c \geqslant 0 \text{ on } \Omega , \text{ is such that }, \\ \exists\, T_1 > 0, \forall\, T > T_1 , \exists\, \gamma_2(T) > 0 \text{ such that} \\ \forall\, (w^0, w^1) \in H_1 \times H , \text{ the solution } w \text{ of (4) satisfies} \\ \int_0^T \int_\Omega c \left| \partial_t w \right|^2 dxdt \geqslant \gamma_2(T) e_1(W)(0) . \end{cases} \qquad (7)$$

The above observability inequalities hold under certain well-known sufficient geometric conditions that we now recall.

2.3 Geometric Assumptions

2.3.1 Multiplier Geometric Conditions

There exist several types of multiplier geometric conditions, depending if the observation domain is located on a part of the boundary or locally distributed within the domain Ω. We recall below, three main conditions, depending on the location of the observability region, or on the number of observations points.

Definition 1 (Boundary geometric condition) We say that a subset $\Gamma_1 \subset \Gamma$ satisfies the boundary multiplier geometric condition, denoted by $(MGC)_{bd}$ (see e.g. [7]), if

$$\exists\, x^0 \in \mathbb{R}^d , \text{ such that } \Gamma_1 \supset \Gamma(x^0) := \{x \in \Gamma \,:\, \nu(x) \cdot (x - x^0) > 0\} . \qquad (8)$$

Remark 1 The point x_0 is called an observation point.

Definition 2 (Internal geometric condition with a single observation point) We say that a subset $\mathcal{O} \subset \Omega$ satisfies the multiplier geometric condition, denoted by $(MGC)_{in}$ (see [7]), if

$$\exists\, x^0 \in \mathbb{R}^d , \text{ such that } \mathcal{O} \text{ contains a neighbourhoord in } \Omega \text{ of } \Gamma(x^0) , \qquad (9)$$

where $\Gamma(x^0)$ is defined in (8).

Definition 3 (Internal geometric condition with multiple observation points)
We say that a subset $\mathcal{O} \subset \Omega$ satisfies the piecewise multiplier geometric condition,
denoted by $(PMGC)_{in}$, if there exist subsets $\Omega_j \subset \Omega$ having Lipschitz boundaries
and points $x_j \in \mathbb{R}^N$, $j = 1, \cdots, J$, such that $\Omega_i \cap \Omega_j = \emptyset$ for $i \neq j$ and $\mathcal{O}$ contains a
neighborhood in Ω of the set $\cup_{j=1}^{J} \gamma_j(x_j) \cup (\Omega \setminus \cup_{j=1}^{J} \Omega_j)$, where

$$\gamma_j(x_j) = \{x \in \partial\Omega_j : (x - x_j) \cdot \nu_j(x) \geq 0\}$$

and ν_j is the outward unit normal to $\partial\Omega_j$.

Remark 2 The points x_j, for $j = 1, \ldots, J$, are called observation points.

Remark 3 It is well-known that (A1) holds for $b = \mathbb{1}_{\Gamma_1}$, when Γ_1 satisfies $(MGC)_{bd}$
and T is sufficiently large (see e.g. [7] for a proof). On the other hand, it is also well-
known that (A2) holds as soon as T is sufficiently large, $c = \mathbb{1}_{\mathcal{O}}$, and $\mathcal{O}$ satisfies
$(MGC)_{in}$ [13] or the refined multiplier geometric condition $(PMGC)$ (see [1, 8, 9]).

2.3.2   Geometric Control Conditions

Definition 4 We say that an open subset ω of Ω satisfies the geometric control
condition, denoted by $(GCC)_{in}$ if there exists a time $T > 0$ such that every
generalized geodesic traveling at speed 1 in Ω meets ω at a time $t < T$.

Definition 5 (Boundary geometric control condition) We say that a subset Γ_1
of the boundary Γ satisfies the boundary geometric control condition, denoted by
$(GCC)_i n$ (see [6]), if every generalized geodesic traveling at speed 1 in Ω meets Γ_1
at a time $t < T$ in a non-diffractive point.

It is well-known that (A1) holds as soon as b is a smooth nonnegative function such
that the set $\Gamma_1 := \{x \in \Gamma : b(x) > 0\}$ and $T > 0$ satisfy $(GCC)_{bd}$ [6], and that this
condition is almost necessary. On the other hand, (A2) holds for T sufficiently large
when c is a smooth nonnegative function such that $\{x \in \Omega : c(x) > 0\}$ and $T > 0$
satisfy $(GCC)_{in}$.

2.4   Observability Results for Homogeneous Systems

We recall the following results proved in [3] (under more general hypotheses).

Theorem 1 *Let $c \in L^{\infty}(\Omega)$. Then the following holds true.*

(a) For all $f \in H^{-1}(\Omega)$ and $g \in L^{2}(\Omega)$ there exists a unique solution (u_1, u_2) of

$$\begin{cases} \partial_t^2 u_1 - \Delta u_1 = 0, & t \in (0, T), x \in \Omega, \\ \partial_t^2 u_2 - \Delta u_2 + c(x)u_1 = 0, & t \in (0, T), x \in \Omega, \\ u_1 = u_2 = 0, & t \in (0, T), x \in \Gamma, \\ u_i(0, \cdot) = 0, & in \ \Omega, i = 1, 2, \\ \partial_t u_1(0, \cdot) = f, \ \partial_t u_2(0, \cdot) = g & in \ \Omega \end{cases} \tag{10}$$

which satisfies

$$\begin{cases} u_1 \in \mathscr{C}([0, T]; L^2(\Omega)) \cap \mathscr{C}^1([0, T]; H^{-1}(\Omega)) \cap \mathscr{C}^2([0, T]; H^{-2}(\Omega)), \\ u_2 \in \mathscr{C}([0, T]; H_0^1(\Omega)) \cap \mathscr{C}^1([0, T]; L^2(\Omega)) \cap \mathscr{C}^2([0, T]; H^{-1}(\Omega)). \end{cases} \tag{11}$$

Moreover, for all $T > 0$, there exists $d_T > 0$ such that the solution of (1) satisfies the following direct inequality

$$\int_0^T \int_\Gamma \left| \frac{\partial u_2}{\partial \nu} \right|^2 d\mathscr{H}_b^{d-1} dt \leqslant d_T \left(|f|_{-1,\Omega}^2 + |g|_{0,\Omega}^2 \right), \ \forall \ (f, g) \in H^{-1}(\Omega) \times L^2(\Omega). \tag{12}$$

(b) Assume (A1), (A2). Then, there exists $T^* > 0$ such that for all $T > T^*$, there exists $c_T > 0$ such that the solution of (10) satisfies the following observability inequality

$$\int_0^T \int_\Gamma \left| \frac{\partial u_2}{\partial \nu} \right|^2 d\mathscr{H}_b^{d-1} dt \geqslant c_T \left(|f|_{-1,\Omega}^2 + |g|_{0,\Omega}^2 \right) \tag{13}$$

for any $(f, g) \in H^{-1}(\Omega) \times L^2(\Omega)$.

Remark 4 Observe that, taking $f \equiv 0$ in (10), we have that $u_1 \equiv 0$ and $u = u_2$ turns out to be the solution of the initial-boundary value problem

$$\begin{cases} \partial_t^2 u - \Delta u = 0, & (t, x) \in (0, T) \times \Omega, \\ u(t, x) = 0, & (t, x) \in (0, T) \times \Gamma, \\ u(0, \cdot) = 0, \ \partial_t u(0, \cdot) = g & in \ \Omega. \end{cases} \tag{14}$$

Consequently, (12) and (13) yield

$$c_T |g|_{0,\Omega}^2 \leqslant \int_0^T \int_\Gamma \left| \frac{\partial u}{\partial \nu} \right|^2 d\mathscr{H}_b^{d-1} dt \leqslant d_T |g|_{0,\Omega}^2 \qquad \forall \ g \in L^2(\Omega). \tag{15}$$

3   Well-Posedness

Proposition 1 *Let $T > 0$ and let $\sigma_1, \sigma_2 : [0, T] \to \mathbb{R}$ be absolutely continuous functions such that*

$$\sigma_1(0) = 1 = \sigma_2(0).$$

Then for all $(f, g) \in H^{-1}(\Omega) \times L^2(\Omega)$, there exists a unique solution (y_1, y_2) of (1) satisfying

$$\begin{cases} y_1 \in \mathscr{C}([0, T]; L^2(\Omega)) \cap \mathscr{C}^1([0, T]; H^{-1}(\Omega)) \cap \mathscr{C}^2([0, T]; H^{-2}(\Omega)), \\ y_2 \in \mathscr{C}([0, T]; H_0^1(\Omega)) \cap \mathscr{C}^1([0, T]; L^2(\Omega)) \cap \mathscr{C}^2([0, T]; H^{-1}(\Omega)). \end{cases} \quad (16)$$

Moreover, (y_1, y_2) is given by

$$\begin{cases} y_1(t, x) = (\sigma_1 * u_1)(t, x) := \displaystyle\int_0^t \sigma_1(s)u_1(t - s, x)ds \\ y_2(t, x) = (\sigma_1 * u_2)(t, x) + ([\sigma_2 - \sigma_1] * w)(t, x) \end{cases} \quad \forall\, (t, x) \in [0, T] \times \Omega, \quad (17)$$

where (u_1, u_2) is the solution of (10) and $w \in \mathscr{C}([0, T]; H_0^1(\Omega)) \cap \mathscr{C}^1([0, T]; L^2(\Omega)) \cap \mathscr{C}^2([0, T]; H^{-1}(\Omega))$ satisfies (14). Furthermore, the following estimate holds true

$$\left\| \partial_t \left(\frac{\partial y_2}{\partial \nu} \right) \right\|_{L^2(0,T; L_b^2(\Gamma))}^2$$

$$\leq 3d_T \left\{ \left(1 + \|\sigma_1'\|_{L^1(0,T)}^2 \right)\left(|f|_{-1,\Omega}^2 + |g|_{0,\Omega}^2\right) + \|\sigma_2' - \sigma_1'\|_{L^1(0,T)}^2 |g|_{0,\Omega}^2 \right\}. \quad (18)$$

Proof Let (y_1, y_2) be defined by (17) and observe that (16) follows from (11), and the fact that σ_i is continuous for $i = 1, 2$. Also, again by (17) we have that $y_i(0, \cdot) = 0$ in Ω for $i = 1, 2$. Furthermore, differentiating y_1 with respect to t we obtain

$$\partial_t y_1(t, x) = \int_0^t \sigma_1(s)\partial_t u_1(t - s, x)ds$$

because $u_1(0, \cdot) = 0$. Hence, $\partial_t y_1(0, \cdot) = 0$. Moreover, recalling (10), we have that

$$\partial_t^2 y_1(t, x) = \sigma_1(t)\partial_t u_1(0, x) + \int_0^t \sigma_1(s)\partial_t^2 u_1(t - s, x)ds$$

$$= \sigma_1(t)f(x) + \int_0^t \sigma_1(s)\Delta u_1(t - s, x)ds = \sigma_1(t)f(x) + \Delta y_1(t, x).$$

Similarly, we compute $\partial_t y_2$ as follows

$$\partial_t y_2(t,x) = \int_0^t \sigma_1(s)\partial_t u_2(t-s,x)ds + \int_0^t [\sigma_2(s) - \sigma_1(s)]\partial_t w(t-s,x)ds, \quad (19)$$

which in turn yields $\partial_t y_2(0,\cdot) = 0$. Moreover,

$$\partial_t^2 y_2(t,x) = \sigma_1(t)\partial_t u_2(0,x) + \int_0^t \sigma_1(s)\partial_t^2 u_2(t-s,x)ds$$

$$+[\sigma_2(t) - \sigma_1(t)]\partial_t w(0,x) + \int_0^t [\sigma_2(s) - \sigma_1(s)]\partial_t^2 w(t-s,x)ds$$

$$= \sigma_1(t)g(x) + \int_0^t \sigma_1(s)\big[\Delta u_2(t-s,x) - c(x)u_1(t-s,x)\big]ds$$

$$+\int_0^t [\sigma_2(s) - \sigma_1(s)]\Delta w(t-s,x)ds$$

$$= \sigma_1(t)g(x) - c(x)(\sigma_1 * u_1)(t,x) + \Delta\Big((\sigma_1 * u_2)(t,x) + \big([\sigma_2 - \sigma_1] * w\big)(t,x)\Big)$$

$$= \sigma_1(t)g(x) - c(x)y_1(t,x) + \Delta y_2(t,x).$$

Therefore, (y_1, y_2) is a solution of (1).

Next, observe that the uniqueness of the solution to (1) is a straightforward consequence of the same property for the wave equation. Then, to complete the proof it suffices to derive estimate (18). For this purpose, observe that, thanks to (12) applied to w, we conclude that

$$\int_0^T \int_\Gamma \left|\frac{\partial w}{\partial \nu}\right|^2 d\mathscr{H}_b^{d-1} dt \leq d_T |g|_{0,\Omega}^2. \quad (20)$$

The above estimate and the analogous one for u_2 imply that $\frac{\partial w}{\partial \nu}$ and $\frac{\partial u_2}{\partial \nu}$ belong to $L^2(0,T;L_b^2(\Gamma))$. Since $\frac{\partial y_2}{\partial \nu} = \sigma_1 * \frac{\partial u_2}{\partial \nu} + [\sigma_2 - \sigma_1] * \frac{\partial w}{\partial \nu}$ and $\sigma_1, \sigma_2 \in H^1(0,T)$, we conclude that $\partial_t\left(\frac{\partial y_2}{\partial \nu}\right) \in L^2(0,T;L_b^2(\Gamma))$. Moreover,

$$\partial_t\left(\frac{\partial y_2}{\partial \nu}\right)(t,x) = \frac{\partial u_2}{\partial \nu}(t,x) + \left(\sigma_1' * \frac{\partial u_2}{\partial \nu}\right)(t,x) + \left([\sigma_2 - \sigma_1]' * \frac{\partial w}{\partial \nu}\right)(t,x). \quad (21)$$

Therefore,

$$\int_0^T \int_\Gamma \left|\partial_t\left(\frac{\partial y_2}{\partial \nu}\right)\right|^2 d\mathscr{H}_b^{d-1} dt$$

$$\leq \int_0^T \int_\Gamma \left|\frac{\partial u_2}{\partial \nu} + \left(\sigma_1' * \frac{\partial u_2}{\partial \nu}\right) + \left([\sigma_2 - \sigma_1]' * \frac{\partial w}{\partial \nu}\right)\right|^2 d\mathscr{H}_b^{d-1} dt$$

$$\leq 3\Big(\int_0^T\!\!\int_\Gamma \Big|\frac{\partial u_2}{\partial \nu}\Big|^2 d\mathcal{H}_b^{d-1}dt + \int_0^T\!\!\int_\Gamma \Big|\sigma_1' * \frac{\partial u_2}{\partial \nu}\Big|^2 d\mathcal{H}_b^{d-1}dt$$

$$+ \int_0^T\!\!\int_\Gamma \Big|[\sigma_2 - \sigma_1]' * \frac{\partial w}{\partial \nu}\Big|^2 d\mathcal{H}_b^{d-1}dt\Big)$$

$$\leq 3\Big(\int_0^T\!\!\int_\Gamma \Big|\frac{\partial u_2}{\partial \nu}\Big|^2 d\mathcal{H}_b^{d-1}dt + \|\sigma_1'\|_{L^1(0,T)}^2 \int_0^T\!\!\int_\Gamma \Big|\frac{\partial u_2}{\partial \nu}\Big|^2 d\mathcal{H}_b^{d-1}dt$$

$$+ \|\sigma_2' - \sigma_1'\|_{L^1(0,T)}^2 \int_0^T\!\!\int_\Gamma \Big|\frac{\partial w}{\partial \nu}\Big|^2 d\mathcal{H}_b^{d-1}dt\Big).$$

The conclusion (18) now follows by inserting estimates (12) and (20) into the above inequality.   □

4   Stability Estimates for the Inverse Source Problem

For any $M, T > 0$ let us set

$$H_M^1(0, T) = \big\{\sigma \in H^1(0, T) : \sigma(0) = 1, \ \|\sigma'\|_{L^2(0,T)} \leq M\big\}. \tag{22}$$

Theorem 2 *Assume* (A1), (A2). *Let* $T > T^*$, *where* $T^* > 0$ *is as in Theorem 1, and* $\sigma_1, \sigma_2 \in H_M^1(0, T)$ *for some* $M > 0$. *Then, for any* $(f, g) \in H^{-1}(\Omega) \times L^2(\Omega)$, *the solution* (y_1, y_2) *of* (1), *in the class* (16), *satisfies*

$$\Big\|\partial_t\Big(\frac{\partial y_2}{\partial \nu}\Big)\Big\|_{L^2(0,T;L_b^2(\Gamma))}^2$$

$$\geq \frac{c_T}{4e^{2TM^2}}|f|_{-1,\Omega}^2 + \Big(\frac{c_T}{4e^{2TM^2}} - d_T\|\sigma_2' - \sigma_1'\|_{L^1(0,T)}^2\Big)|g|_{0,\Omega}^2 \tag{23}$$

with c_T *and* d_T *given by Theorem 1.*

Before proving Theorem 2, we derive a simple estimate for solutions of integral equations.

Lemma 1 *Let* $(X, \|\cdot\|)$ *be a Banach space and let* $r \in L^2(0, T)$. *Then, for any* $h \in L^2(0, T : X)$ *the solution* $z \in L^2(0, T; X)$ *of*

$$z(t) + (r * z)(t) = h(t) \qquad t \in [0, T] \ a.e.$$

satisfies

$$\int_0^T \|z(t)\|^2 \, dt \leq 2e^{2T\|r\|_{L^2(0,T)}^2} \int_0^T \|h(t)\|^2 \, dt. \tag{24}$$

Proof Since $z = h - r * z$, we have

$$\|z(t)\|^2 \leqslant 2\|h(t)\|^2 + 2\|(r * z)(t)\|^2 \leqslant 2\|h(t)\|^2 + 2\|r\|_{L^2(0,T)}^2 \int_0^t \|z(s)\|^2 ds.$$

The conclusion follows by the Gronwall lemma.   □

Proof of Theorem 2 Observe that, in view of (21),

$$\frac{\partial u_2}{\partial \nu}(t, x) + \left(\sigma_1' * \frac{\partial u_2}{\partial \nu}\right)(t, x) = \partial_t \left(\frac{\partial y_2}{\partial \nu}\right)(t, x) + \left([\sigma_1 - \sigma_2]' * \frac{\partial w}{\partial \nu}\right)(t, x)$$

with $X = L^2(0, T; L_b^2(\Gamma))$. By (24) we have

$$\int_0^T \int_\Gamma \left|\frac{\partial u_2}{\partial \nu}\right|^2 d\mathcal{H}_b^{d-1} dt$$

$$\leqslant 2e^{2TM^2} \int_0^T \int_\Gamma \left|\partial_t \left(\frac{\partial y_2}{\partial \nu}\right) + \left([\sigma_1 - \sigma_2]' * \frac{\partial w}{\partial \nu}\right)\right|^2 d\mathcal{H}_b^{d-1} dt.$$

Therefore,

$$\int_0^T \int_\Gamma \left|\frac{\partial u_2}{\partial \nu}\right|^2 d\mathcal{H}_b^{d-1} dt \leqslant 4e^{2TM^2} \left\{ \int_0^T \int_\Gamma \left|\partial_t \left(\frac{\partial y_2}{\partial \nu}\right)\right|^2 d\mathcal{H}_b^{d-1} dt \right.$$

$$\left. + \|\sigma_2' - \sigma_1'\|_{L^1(0,T)}^2 \int_0^T \int_\Gamma \left|\frac{\partial w}{\partial \nu}\right|^2 d\mathcal{H}_b^{d-1} dt \right\}$$

$$\leqslant 4e^{2TM^2} \left\{ \int_0^T \int_\Gamma \left|\partial_t \left(\frac{\partial y_2}{\partial \nu}\right)\right|^2 d\mathcal{H}_b^{d-1} dt + d_T \|\sigma_2' - \sigma_1'\|_{L^1(0,T)}^2 |g|_{0,\Omega}^2 \right\}$$

owing to the direct inequality in (15). Hence, appealing to (13) to bound the term on the left side of the above inequality from below, we obtain

$$c_T (|f|_{-1,\Omega}^2 + |g|_{0,\Omega}^2)$$

$$\leqslant 4e^{2TM^2} \left\{ \int_0^T \int_\Gamma \left|\partial_t \left(\frac{\partial y_2}{\partial \nu}\right)\right|^2 d\mathcal{H}_b^{d-1} dt + d_T \|\sigma_2' - \sigma_1'\|_{L^1(0,T)}^2 |g|_{0,\Omega}^2 \right\}$$

or

$$\frac{c_T}{4e^{2TM^2}} |f|_{-1,\Omega}^2 + \left(\frac{c_T}{4e^{2TM^2}} - d_T \|\sigma_2' - \sigma_1'\|_{L^1(0,T)}^2\right) |g|_{0,\Omega}^2$$

$$\leqslant \int_0^T \int_\Gamma \left|\partial_t \left(\frac{\partial y_2}{\partial \nu}\right)\right|^2 d\mathcal{H}_b^{d-1} dt,$$

which completes the proof.   □

The following is an immediate corollary of the above result.

Theorem 3 *Assume (A1), (A2). and let $T > T^*$, where $T^* > 0$ is as in Theorem 1. Let $\sigma_1, \sigma_2 \in H^1_M(0, T)$ for some $M > 0$ and fix $\kappa \in \mathbb{R}$ such that*

$$0 < \kappa < \frac{e^{-TM^2}}{2} \sqrt{\frac{c_T}{d_T}}, \qquad (25)$$

with c_T and d_T given by Theorem 1. If $\|\sigma'_2 - \sigma'_1\|_{L^1(0,T)} \leq \kappa$, then, for any $(f, g) \in H^{-1}(\Omega) \times L^2(\Omega)$, the solution (y_1, y_2) of (1), in the class (16), satisfies

$$|f|^2_{-1,\Omega} + |g|^2_{0,\Omega} \leq \frac{4e^{2TM^2}}{c_T - 4d_T\kappa^2 e^{2TM^2}} \left\| \partial_t \left(\frac{\partial y_2}{\partial \nu} \right) \right\|^2_{L^2(0,T;L^2_b(\Gamma))}. \qquad (26)$$

Remark 5 In all the above results, the requirement that $\sigma_1(0) = 1 = \sigma_2(0)$ can be easily removed requiring that $|\sigma_i(0)| \geq \rho > 0$ for $i = 1, 2$. Indeed, applying Theorem 3 with $\tilde{\sigma}_i(t) = \sigma_i(t)/\sigma_i(0)$ for $i = 1, 2, \tilde{f}(x) = \sigma_1(0)f(x)$, and $\tilde{g}(x) = \sigma_2(0)g(x)$ we conclude that, if

$$\|\sigma'_i\|_{L^2(0,T)} \leq M\rho \quad \text{for } i = 1, 2 \quad \text{and} \quad \left\| \frac{\sigma'_2}{\sigma_2(0)} - \frac{\sigma'_1}{\sigma_1(0)} \right\|_{L^1(0,T)} \leq \kappa$$

with κ as in (25), then

$$\rho^2 \left(|f|^2_{-1,\Omega} + |g|^2_{0,\Omega} \right) \leq \frac{4e^{2TM^2}}{c_T - 4d_T\kappa^2 e^{2TM^2}} \left\| \partial_t \left(\frac{\partial y_2}{\partial \nu} \right) \right\|^2_{L^2(0,T;L^2_b(\Gamma))}.$$

Cleaner estimates can be obtained when the derivative of at least one of the σ_i's is supposed to be a positive definite kernel. Given $h \in L^1_{loc}(0, \infty)$, we recall that h is a *positive definite kernel* if

$$\int_0^t \langle h * y(s), y(s) \rangle \, ds \geq 0, \qquad t \geq 0, \qquad (27)$$

for any $y \in L^2_{loc}(0, \infty; X)$. For $h \in L^1(0, \infty)$ it is well known that h is a positive definite kernel if and only if $\operatorname{Re} \hat{h}(i\omega) \geq 0$ for any $\omega \in \mathbb{R}$ (see [10, Theorem 2]), where $\hat{h}$ denotes the Laplace transform of h, that is

$$\hat{h}(z) := \int_0^\infty e^{-zt} h(t) \, dt, \qquad z \in \mathbb{C}.$$

The proof of Theorem 2 simplifies under the above positivity assumption, yielding the following result.

Proposition 2 *Assume* (A1), (A2). *and let* $T > T^*$, *where* $T^* > 0$ *is as in Theorem 1. Let* $\sigma_1, \sigma_2 \in H^1(0, T)$ *be such that* $\sigma_1(0) = 1 = \sigma_2(0)$ *and suppose* σ_1', *extended as zero outside* $[0, T]$, *is a positive definite kernel. Fix* $\kappa \in \mathbb{R}$ *such that*

$$0 < \kappa < \frac{1}{2}\sqrt{\frac{c_T}{d_T}}, \tag{28}$$

with c_T *and* d_T *given by Theorem 1. If* $\|\sigma_2' - \sigma_1'\|_{L^1(0,T)} \le \kappa$, *then, for any* $(f, g) \in H^{-1}(\Omega) \times L^2(\Omega)$, *the solution* (y_1, y_2) *of* (1), *in the class* (16), *satisfies*

$$|f|_{-1,\Omega}^2 + |g|_{0,\Omega}^2 \le \frac{2}{c_T - 2d_T\kappa^2}\left\|\partial_t\left(\frac{\partial y_2}{\partial \nu}\right)\right\|_{L^2(0,T;L_b^2(\Gamma))}^2. \tag{29}$$

A similar result holds true when σ_2', *extended as zero outside* $[0, T]$, *is a positive definite kernel.*

5 A Negative Result

The above results concern cases for which c keeps a constant sign within the spatial domain. If that is not the case, one can exhibit counterexamples to unique continuation for the system (1) (see [5]). We shall apply these results to show that identification may also fail to hold for the source reconstruction problem.

We consider the following one-dimensional wave system.

$$\begin{cases} y_{1,tt} - y_{1,xx} = \sigma_1(t)f(x), & 0 < t < T, 0 < x < \pi, \\ y_{2,tt} - y_{2,xx} + c(x)y_1 = \sigma_2(t)g(x), & 0 < t < T, 0 < x < \pi, \\ y_i(t, 0) = y_i(t, \pi) = 0, & \text{for } i = 1, 2, 0 < t < T, \\ y_i(0, x) = 0, y_{i,t}(0, x) = 0, & \text{for } i = 1, 2, 0 < x < \pi. \end{cases} \tag{30}$$

The corresponding homogeneous system is now given by

$$\begin{cases} u_{1,tt} - u_{1,xx} = 0, & 0 < t < T, 0 < x < \pi, \\ u_{2,tt} - u_{2,xx} + c(x)u_1 = 0, & 0 < t < T, 0 < x < \pi, \\ u_i(t, 0) = u_i(t, \pi) = 0, & \text{for } i = 1, 2, 0 < t < T, \\ u_i(0, x) = 0, & \text{for } i = 1, 2, 0 < x < \pi, \\ u_{1,t}(0, x) = f(x), u_{2,t}(0, x) = g(x), & 0 < x < \pi. \end{cases} \tag{31}$$

Theorem 4 *Assume that $\sigma_1 \equiv \sigma_2$. Assume moreover that the coupling coefficient is given by*

$$c(x) = \cos(m_0 x), \ x \in (0, \pi), \ where \ m_0 = 2p_0 + 1, \ p_0 \in \mathbb{N}. \tag{32}$$

Let $\mathcal{N}$ to be the set of all pairs (f, g) of functions defined on $(0, \pi)$ such that

$$f(x) = \sum_{k=1}^{\infty} k b_k \sin(kx)$$

$$g(x) = \sum_{k=1}^{m_0-1} \left(\frac{k b_k}{4k^2 - m_0^2} + \frac{(k + m_0) b_{m_0+k}}{2m_0(2k + m_0)} - \frac{(m_0 - k) b_{m_0-k}}{2m_0(m_0 - 2k)} \right) \sin(kx)$$

$$+ \sum_{k=m_0+1}^{\infty} \left(\frac{k b_k}{4k^2 - m_0^2} + \frac{(k + m_0) b_{m_0+k}}{2m_0(2k + m_0)} - \frac{(k - m_0) b_{k-m_0}}{2m_0(2k - m_0)} \right) \sin(kx)$$

$$+ \left(b_{m_0} + b_{2m_0} \right) \frac{\sin(m_0 x)}{3m_0}$$

with

$$\sum_{k=1}^{\infty} b_k^2 < \infty.$$

Then $\mathcal{N}$ is an infinite-dimensional subspace of $H^{-1}(0, \pi) \times L^2(0, \pi)$. Moreover, the solution of (30) satisfies

$$\partial_t \left(\frac{\partial y_2}{\partial x} \right) (., 0) \equiv 0 \ on \ (0, T). \tag{33}$$

if and only if $(f, g) \in \mathcal{N}$.

Remark 6 Hence $\mathcal{N} \setminus \{(0, 0, 0, 0)\}$ is exactly the set of initial data for which identification does not hold.

Proof Since $\sigma_1 = \sigma_2$, we have, thanks to (21),

$$\partial_t \left(\frac{\partial y_2}{\partial x} \right) (t, x) = \frac{\partial u_2}{\partial x} (t, x) + \left(\sigma_1' * \frac{\partial u_2}{\partial x} \right) (t, x),$$

where (u_1, u_2) is the solution of (31). Hence,

$$\partial_t \left(\frac{\partial y_2}{\partial x} \right) (t, 0) \equiv 0,$$

if and only if

$$\frac{\partial u_2}{\partial x}(t,0) \equiv 0.$$

We now use the following result for the solutions of the more general problem

$$\begin{cases} u_{1,tt} - u_{1,xx} = 0, & 0 < t < T, 0 < x < \pi, \\ u_{2,tt} - u_{2,xx} + c(x)u_1 = 0, & 0 < t < T, 0 < x < \pi, \\ u_i(t,0) = u_i(t,\pi) = 0, & \text{for } i = 1,2, 0 < t < T, \\ u_i(0,x) = 0, & \text{for } i = 1,2, 0 < x < \pi, \\ u_i(0,x) = u_i^0(x), u_{i,t}(0,x) = u_i^1(x), & \text{for } i = 1,2, 0 < x < \pi. \end{cases} \tag{34}$$

Theorem 5 (Theorem 5.1 [5]) *Assume that the coupling coefficient is given by* (32) *Define*

$$\mathscr{I} = \left\{ (u_1^0, u_2^0, u_1^1, u_2^1), u_1^0(x) = \sum_{k=1}^{\infty} a_k \sin(kx), u_1^1(x) = \sum_{k=1}^{\infty} kb_k \sin(kx), \right.$$

$$u_2^0(x) = \sum_{k=1}^{m_0-1} \left(\frac{a_k}{4k^2 - m_0^2} + \frac{a_{m_0+k}}{2m_0(2k+m_0)} - \frac{a_{m_0-k}}{2m_0(m_0-2k)} \right) \sin(kx)$$

$$+ \sum_{k=m_0+1}^{\infty} \left(\frac{a_k}{4k^2 - m_0^2} + \frac{a_{m_0+k}}{2m_0(2k+m_0)} - \frac{a_{k-m_0}}{2m_0(2k-m_0)} \right) \sin(kx)$$

$$+ \left(a_{m_0} + \frac{a_{2m_0}}{2} \right) \frac{\sin(m_0 x)}{3m_0^2}, \quad x \in (0,\pi),$$

$$u_2^1(x) = \sum_{k=1}^{m_0-1} \left(\frac{kb_k}{4k^2 - m_0^2} + \frac{(k+m_0)b_{m_0+k}}{2m_0(2k+m_0)} - \frac{(m_0-k)b_{m_0-k}}{2m_0(m_0-2k)} \right) \sin(kx)$$

$$+ \sum_{k=m_0+1}^{\infty} \left(\frac{kb_k}{4k^2 - m_0^2} + \frac{(k+m_0)b_{m_0+k}}{2m_0(2k+m_0)} - \frac{(k-m_0)b_{k-m_0}}{2m_0(2k-m_0)} \right) \sin(kx)$$

$$+ \left(b_{m_0} + b_{2m_0} \right) \frac{\sin(m_0 x)}{3m_0}, \quad x \in (0,\pi),$$

$$\left. \sum_{k=1}^{\infty} a_k^2 < \infty, \sum_{k=1}^{\infty} b_k^2 < \infty \right\}. \tag{35}$$

Then $\mathscr{I}$ is a set of infinite dimension such that

$$\mathscr{I} \subset L^2(0,\pi) \times H_0^1(0,\pi) \times H^{-1}(0,\pi) \times L^2(0,\pi),$$

and for all $(u_1^0, u_2^0, u_1^1, u_2^2) \in \mathscr{I}$, *the solution of* (34) *satisfies*

$$\frac{\partial u_2}{\partial x}(., 0) \equiv 0 \; on \; (0, T) \, . \tag{36}$$

Moreover $\mathscr{I} \setminus \{(0, 0, 0, 0)\}$ *is the set of initial data for which unique continuation does not hold.*

Noting that $(f, g) \in \mathscr{N}$ is equivalent to $(0, 0, u_1^1, u_2^1) \in \mathscr{I}$, where $u_1^1 = f$, $u_2^1 = g$, we easily conclude. $\qquad \Box$

Acknowledgements This research is partially supported by the INdAM National Group GNAMPA. This work was completed while the first author was visiting the Institut Henri Poincaré and Institut des Hautes Études Scientifiques on a senior CARMIN position.

References

1. Alabau-Boussouira, F.: Convexity and weighted integral inequalities for energy decay rates of nonlinear dissipative hyperbolic systems. Appl. Math. Optim. **51**, 61–105 (2005)
2. Alabau-Boussouira, F.: Controllability of cascade coupled systems of multi-dimensional evolution PDE's by a reduced number of controls. C. R. Acad. Sci. Paris Sér. I **350**, 577–582 (2012)
3. Alabau-Boussouira, F.: Insensitizing controls for the scalar wave equation and exact controllability of 2-coupled cascade systems of PDE's by a single control. Math. Control Signals Syst. **26**, 1–46 (2014)
4. Alabau-Boussouira, F.: A hierarchic multi-level energy method for the control of bi-diagonal and mixed n-coupled cascade systems of PDE's by a reduced number of controls. Adv. Differ. Equ. **18**, 1005–1072 (2013)
5. Alabau-Boussouira, F.: On the influence of the coupling on the dynamics of single-observed cascade systems of PDE's. Math. Control Relat. Fields **5**, 1–30 (2015)
6. Bardos, C., Lebeau, G., Rauch, J.: Sharp sufficient conditions for the observation, control, and stabilization of waves from the boundary. SIAM J. Control Optim. **30**, 1024–1065 (1992)
7. Lions, J.-L.: Contrôlabilité exacte et stabilisation de systèmes distribués. Volume 1 of Collection RMA, Masson, Paris (1988)
8. Liu K.: Locally distributed control and damping for the conservative systems. SIAM J. Control Optim. **35**, 1574–1590 (1997)
9. Martinez, P.: A new method to obtain decay rate estimates for dissipative systems with localized damping. Rev. Math. Comput. **1**, 251–283 (1999)
10. Nohel, J.A., Shea D.F.: Frequency domain methods for Volterra equations. Adv. Math. **22**, 278–304 (1976)
11. Puel, J.-P., Yamamoto, M.: On a global estimate in a linear inverse hyperbolic problem. Inverse Probl. **12**, 995–1002 (1996)
12. Yamamoto, M.: Stability, reconstruction formula and regularization for an inverse source hyperbolic problem by a control method. Inverse Probl. **11**, 481–496 (1995)
13. Zuazua, E.: Exponential decay for the semilinear wave equation with locally distributed damping. Commun. Partial Differ. Equ. **15**, 205–235 (1990)

Remarks on Stochastic Navier-Stokes Equations

Franco Flandoli

Abstract A theory of stochastic Navier-Stokes equations has been developed in recent years. A few basic facts and open problems are recalled, stressing the role and difference between the model with additive and with multiplicative transport noise. In view of the practical issue of numerical simulation of expected values, a few remarks on Girsanov transform and Kolmogorov equation are given.

Keywords Stochastic Navier-Stokes equations • Additive noise • Transport noise • Girsanov transform • Kolmogorov equations

1 Introduction

Stochastic Navier-Stokes equations have become a standard topic in the community of people working on Stochastic Partial Differential Equations (SPDEs); since the number of works is too large, let us quote only a few books and review, [1, 7, 13, 14, 21, 24, 37].

From the viewpoint of applications, stochastic Navier-Stokes equations and other SPDE models of fluid dynamics have been used for instance in the investigations of turbulence, both numerically and from the viewpoint of theoretical physics, see the review [25]. They look promising for climate research too, for instance as simplified models where it is convenient to model certain degrees of freedom (secondary or less known variables, small scale variables) by noise; or as models which may allow to estimate uncertainty of predictions.

The purpose of this note is a very short review of a few instances of problems concerning stochastic Navier-Stokes equations, possibly of interest for climate applications. After a very short remind of the mathematical set-up and basic results of the deterministic mathematical theory, we discuss two paradigmatic form of noise which could be introduced in the Navier-Stokes equations: additive noise, transport noise. For both, we indicate one feature, related to energy balance laws, which may help to decide which noise could be used in a specific example. Much work should

F. Flandoli (✉)

Dipartimento di Matematica, Università di Pisa, Largo Bruno Pontecorvo 5, I–56127 Pisa, Italia

e-mail: flandoli@dma.unipi.it

© Springer International Publishing Switzerland 2016   51

F. Ancona et al. (eds.), *Mathematical Paradigms of Climate Science*, Springer

INdAM Series 15, DOI 10.1007/978-3-319-39092-5_4

be done here to understand, for instance, if backscatter (or other typical questions of applied disciplines, notably climatology) may be modeled by noise taking advantage of the understanding of its energy balance laws.

Then, we shortly recall a few mathematical results, attempts and open problems concerning the foundational question of well posedness in 3D. This topic would require much more space and we address to some more extended discussions in the literature, like [13, 18, 30].

Finally, we present two potential lines of research having to do with the quantitative approximation of solutions and their statistics. Opposite to direct Monte Carlo simulations, two perturbation methods based on Girsanov transform and Kolmogorov equations are discussed. Again, this is only a beginning which should be explored in more detail.

2   Preliminary: The Deterministic Navier-Stokes Equations

The system of Navier-Stokes equations has the form

$$\frac{\partial u}{\partial t} + u \cdot \nabla u + \nabla p = \nu \Delta u + f, \qquad u|_{t=0} = u_0$$

$$\operatorname{div} u = 0$$

$t \in [0, T]$, $x \in D \subset \mathbb{R}^d$, $d = 2, 3$, where $u = u(x, t)$ is the velocity field, $p = p(x, t)$ is the pressure field, $f = f(x, t)$ is the body force, ν is the viscosity. The open set D is the domain occupied by the fluid and boundary conditions have to be imposed; for instance, for theoretical or numerical investigations, the torus $D = [0, L]^d$ is often considered and periodic boundary conditions are assumed. Without entering in details not needed here (see for instance [26]), let us only introduce the spaces $\mathscr{D} = \{v \in C^{\infty}(D), \operatorname{div} v = 0, v \text{periodic}\}$, $H = $ closure of $\mathscr{D}$ in $L^2(D; \mathbb{R}^d)$; $V = $ closure of $\mathscr{D}$ in $W^{1,2}(D; \mathbb{R}^d)$; V' is the dual of V.

Theorem 1 (existence of weak solutions) *Given $u_0 \in H$, $f \in L^2(0, T; V')$ there exists a function $u \in L^{\infty}(0, T; H) \cap L^2(0, T; V)$, weakly continuous in H, which satisfies the Navier-Stokes equations in the sense of distributions and such that*

$$\int_D |u(t, x)|^2\, dx + 2\nu \int_0^t \int_D |\nabla u(s, x)|^2\, dxds \leq \int_D |u_0(x)|^2\, dx$$

$$+ \int_0^t \int_D u \cdot f dxds.$$

Theorem 2 (local regular solutions) *If $d = 3$, $u_0 \in V$, $f \in L^2(0, T; H)$, there is a unique solution, continuous in V, on some interval $[0, \tau)$, $\tau > 0$.*

Theorem 3 (well posedness in 2D) *If $d = 2$, the weak solution is unique. If $u_0 \in V, f \in L^2 (0, T; H)$, the solution is continuous in V globally in time.*

See for instance [26] for these and other results. The main foundational open problems (see [9] for the formulation of the millennium prize problem) in $d = 3$ are *uniqueness of weak solutions* and *global existence of regular solutions*.

Beside these, there are several other open problems, concerning the inviscid case ($v = 0$, Euler equations), also in $d = 2$, the inviscid limit, statistical properties for very small v (turbulence regime), etc.

3 Stochastic Perturbations

For specific practical purposes, several random perturbations could be of interest (e.g. noise on the boundary). Here we discuss only two of them which in a sense are the main paradigms:

- *additive noise ξ*

$$\frac{\partial u}{\partial t} + u \cdot \nabla u + \nabla p - v\Delta u = \xi,$$

- *transport type noise*

$$\frac{\partial u}{\partial t} + u \cdot \nabla u + \nabla p - v\Delta u = \xi \cdot \nabla u.$$

The noise $\xi = \xi (x, t)$ will be white (delta-correlated) in time, with some degree of correlation in space. In the usual language of applied sciences,

$$\left\langle \xi_\alpha (x, t) \, \xi_\beta (y, s) \right\rangle = \delta (t - s) \, Q_{\alpha\beta} (x, y) \qquad \alpha, \beta = 1, \ldots, d$$

where $Q (x, y)$ is the matrix-valued space-covariance of the noise and $\langle \cdot \rangle$ denotes the average. We shall always assume $Q (x, y) = Q (x - y)$, equivalent to the space-homogeneity of the random field ξ (the law is invariant by space-translations). Moreover we assume that $\operatorname{div} \xi = 0$.

A rigorous mathematical model of ξ is given by a probability space $(\Omega, \mathscr{F}, P)$, a sequence of independent Brownian motions $W_k (t)$ on $(\Omega, \mathscr{F}, P)$, $k = 1, 2, \ldots$, a sequence of divergence free vector fields $\sigma^k : D \to \mathbb{R}^d$ such that

$$Q_{\alpha\beta} (x, y) = \sum_{k=1}^\infty \sigma_\alpha^k (x) \, \sigma_\beta^k (y).$$

Assumptions of summability and regularity of σ^k are required, depending on the result. With these data, the time-distribution

$$\xi(x, t) = \sum_{k=1}^{\infty} \sigma^k(x) \frac{dW_k(t)}{dt}$$

is a white noise in time, with covariance $Q(x, y)$ in space. With these notations, the two stochastic Navier-Stokes equations introduced so far are

$$du + (u \cdot \nabla u + \nabla p - \nu \Delta u) \, dt = \sum_{k=1}^{\infty} \sigma^k(x) \, dW_k(t)$$

$$du + (u \cdot \nabla u + \nabla p - \nu \Delta u) \, dt = \sum_{k=1}^{\infty} \sigma^k(x) \cdot \nabla u \circ dW_k(t)$$

(additive and transport type noise, respectively).

For rigorous investigations, both equations must be interpreted weakly (integrated in time and against smooth test functions), similarly to the deterministic case (we omit this detail, see for instance [15]). In the case of transport noise, the notation $\circ$ stands for Stratonovich stochastic integration; see for instance [16]; Stratonovich integration is the natural one from the physical viewpoint, as a limit of noise regularizations (Wong-Zakai principle, see the Appendix of [16] for an example of rigorous result).

A question of interest for applications is: which noise pertains to a specific problem? In absence of detailed knowledge of the statistics of the random perturbation, it may be useful to have general ideas. Let us give two of them.

3.1 Motivations for Additive Noise

Additive noise is the generic random forcing assumed in most investigations. It has a distinctive property: *the average energy injected into the system is a priori known.*

Indeed, by Itô formula (used here in heuristic way; $E[\cdot]$ denotes mathematical expectation, corresponding to $\langle \cdot \rangle$ of the applied literature)

$$\frac{1}{2} \frac{\partial}{\partial t} E\left[|u|^2\right] + E[(u \cdot \nabla u + \nabla p - \nu \Delta u) \cdot u] = \frac{1}{2} Trace(Q(0)).$$

This is a (local in space-time) energy balance. From this we see that $\frac{1}{2} Trace(Q(0))$ is the *rate of average energy input per unit volume and unit time*. It is constant in space because we have assumed space-homogeneity of the random field, $Q(x, y) = Q(x - y)$. Notice also the integrated version (global in space-time) of this energy identity, which holds under usual boundary conditions (rigorously in 3D one can prove only the existence of solutions satisfying the energy inequality, see Theorem 4 below), is

$$\frac{1}{2}E\left[\int_D |u(t,x)|^2\,dx\right] + \nu E\left[\int_0^T \int_D |\nabla u(s,x)|^2\,dxds\right]$$
$$= \frac{1}{2}E\left[\int_D |u(0,x)|^2\,dx\right] + \frac{1}{2}Trace(Q(0))T|D|$$

which identifies $\frac{1}{2}Trace(Q(0))T|D|$ as the global energy input by the noise.

In contrast, in the case of a deterministic force f, we (formally) have

$$\frac{1}{2}\frac{\partial}{\partial t}|u|^2 + (u\cdot\nabla u + \nabla p - \nu\Delta u)\cdot u = u\cdot f.$$

Here the energy input $u\cdot f$ depends on the unknown solution; in principle, we could have $u\cdot f = 0$, or $\int_0^T\int_D u\cdot f dxds = 0$, even for non zero force. Thus the energy injected by a deterministic force is not known a priori, it is not under control.

Due to this motivation, additive noise is often used in turbulence investigation: one has energy balance under control.

3.2 Motivations for Transport Noise

Recall that the transport noise, in Stratonovich form, is the term $\xi \circ \nabla u$ which adds to the other terms of the Navier-Stokes equations, in particular to the term $u\cdot\nabla u$. Its meaning is thus of random fluctuations *superimposed to the velocity field*, at the Lagrangian level: $u \longrightarrow u + \xi$ (the velocity field u in the transport term $u\cdot\nabla$ is replaced by $u + \xi$).

This model is (formally) *energy neutral*: no contribution to the *global* energy balance. By (formal) Itô formula (in Stratonovich form; this is essential) we get

$$\frac{1}{2}\frac{\partial}{\partial t}|u|^2 + (u\cdot\nabla u + \nabla p - \nu\Delta u)\cdot u = (\xi \circ \nabla u)\cdot u$$

and then, integrating over the full domain D (with suitable boundary conditions) we get

$$\frac{1}{2}\int_D |u(t,x)|^2\,dx + \nu\int_0^t \int_D |\nabla u(s,x)|^2\,dxds = \frac{1}{2}\int_D |u_0(x)|^2\,dx.$$

The noise disappears.

To be more cautious, we should keep in mind the possibility that this kind of noise, due to the fact that it enforces cascade of energy, contributes to induce *anomalous* dissipation of energy (very fast cascade of energy from large to small scale, so fast that energy disappears "at infinity" in wave space). This is not proved or observed until now for 3D Navier-Stokes equations, so it is only a theoretical possibility (recall that the energy identity above is not known, only existence of solutions satisfying energy inequalities is known), but it has been proved for very simplified models, called dyadic models of turbulence, see [3, 4].

4   Remarks on Well-Posedness

In very rough terms we can say that the theory of *stochastic* Navier-Stokes equations has the same theorems of existence, uniqueness, regularity and the *same open problems* as the deterministic case.

The similarity is affected by details. Recall that a theorem of existence is called *strong* (in the probabilistic sense) if the solution is adapted to the completed filtration of the noise, *weak* otherwise. For stochastic 3D Navier-Stokes equations only weak existence theorems are known (strong in 2D). The basic example of existence theorem, generalizing Theorem 1 to the stochastic case, is (see [15]):

Theorem 4 (weak existence of weak solutions, additive noise) *Given $u_0 \in H$, there exists a probability space $(\Omega, \mathscr{F}, P)$, a sequence of independent Brownian motions $W_k(t)$ on $(\Omega, \mathscr{F}, P)$, $k = 1, 2, \ldots$, and a stochastic process u, weakly continuous in H, which satisfies the stochastic 3D Navier-Stokes equations in weak form and such that*

$$E\left[\int_D |u(t,x)|^2 \, dx\right] + 2\nu E\left[\int_0^t \int_D |\nabla u(s,x)|^2 \, dxds\right]$$
$$\leq \int_D |u_0(x)|^2 \, dx + Trace(Q(0))t \, |D| \, .$$

A fascinating research direction, which occupied the efforts of a number of people for several years (at least the last 15 years) has been the attempt to improve the deterministic theory thanks to noise. Unfortunately, only very partial progresses have been reached and it is no more so clear if it is reasonable to expect such *regularization by noise*.

The reason to try was due to the analogy with the following well known fact in finite dimension: noise improves uniqueness. The solution to the equation in $\mathbb{R}^d$

$$dX(t) = b(X(t)) \, dt + dW(t), \qquad X(0) = x_0 \in \mathbb{R}^d$$

is (pathwise) unique when $b \in L^\infty$, see [36] or even when $b \in L^p$ for $p > d$, see [23].

Is it possible to generalize this result to infinite dimensions? Recently we have proved results for abstract evolution equations

$$du(t) = Au(t) dt + B(u(t)) dt + dW(t) \tag{1}$$

where $A = A^* < 0$: see [31] when B is Hölder and bounded, [32] when B is L^∞. But they do not cover relevant fluid dynamics equations, first of all because B must be at least locally bounded on H.

The main contribution in the case of stochastic 3D Navier-Stokes equations has been the theory of *Markov selections with Strong Feller property*, developed by [8, 17, 18, 30, 35], see [13] for a review. Let us explain in a few words the result. From Theorem 4, given $u_0 \in H$, we could have more than one weak solution, but at least one exists. By "selection" we mean that we associate, to each $u_0 \in H$, a specific solution $u(t; u_0)$, possibly with special properties. For instance, it is not too difficult to prove the existence of Borel measurable selections, namely selections which depend measurably on u_0. Much more difficult is to prove the existence of Markov selections, namely selections $u_0 \longmapsto u(t; u_0)$ such that the family of maps P_t defined as

$$P_t \varphi(u_0) := E[\varphi(u(t; u_0))], \qquad \varphi \in B_b(H)$$

($B_b(H)$ is the space of Borel bounded functions on H) satisfies the semigroup property

$$P_t P_s \varphi = P_{t+s} \varphi. \tag{2}$$

In general, without any particular assumption on the noise (thus even without noise), a Markov selection exists (identity (2) holds only a.s. in time), see [18]. But the main result, of regularization by noise, is:

Theorem 5 *When the noise satisfies suitable assumptions of non-degeneracy, each Markov selection is strong Feller, with respect to variations of u_0 in V.*

Strong Feller means that the map

$$u_0 \longmapsto P_t \varphi(u_0) = E[\varphi(u(t; u_0))], \qquad \varphi \in B_b(H)$$

is continuous in the V-topology. The law of the solution $u(t; u_0)$ depends continuously (in total variation!) on $u_0 \in V$. No result of continuous dependence on initial conditions, for arbitrarily large time, is known in the deterministic case! Intuitively, this continuity result looks so close to uniqueness that it gave the impression to be close to solve the main open problem of uniqueness. However, after a few more years, it looks clear that the Strong Feller property is fully disjoint from any uniqueness statement.

Concerning the previous result, recall Theorem 2: locally in time, we have seen that V-regularity is maintained; similarly one can prove that, for short time, solutions depend continuously on u_0 in the V-topology. Thus a claim like the one of Theorem 5 on small time would just be a generalization of the deterministic theory. The novelty is that continuous dependence holds for arbitrary large time. This is due to a careful use of Markov property and to the fact that continuous dependence is in total variation. A sort of principle is identified: for a PDE without uniqueness, under suitable non-degeneracy assumptions on the noise, continuous dependence of a selection $u(t; u_0)$ on u_0 may hold if we have two ingredients, local-in-time well posedness + Markov property.

Until now we have discussed the case of additive noise only. For transport type noise there are existence results similar to Theorem 4 above, see [15, 27]. Concerning regularization by noise, transport noise is very promising for inviscid models; see [2, 4, 16, 16] where uniqueness is restored by noise for linear transport equations with rough coefficients and for nonlinear dyadic type models.

5 Practical Issues

We advise the reader that this section is intentionally written in heuristic way and it has only an explorative character; any rigorous result has still to be done. Some basic tools to handle rigorously these topics can be found for instance in the books [33, 34].

Consider a stochastic differential equation (e.g. Navier-Stokes equations) in a Hilbert space H of the form

$$du(t) = Au(t)\, dt + B(u(t))\, dt + \sqrt{Q}dW(t), \qquad u(0) = u_0 \in H$$

where $A : D(A) \subset H \to H$ is a linear operator (usually the Laplacian, in applications), B is a nonlinear operator (usually the transport terms in fluid dynamics) and $\sqrt{Q}dW(t)$ is the noise (we discuss only additive noise to fix the ideas). The notations we use here for the noise can be understood as follows. One has, on a probability space, a sequence of independent Brownian motions $W_k(t)$, $k \in \mathbb{N}$ (as in the previous sections); moreover, one has, on the Hilbert space H, a non-negative selfadjoint bounded operator Q, and a complete orthonormal system $\{e_k\}_{k\in\mathbb{N}}$. Then $\sqrt{Q}dW(t)$ denotes the formal random series $\sum_{k=1}^{\infty} \sqrt{Q}e_k dW_k(t)$ (the series converges in suitable distributional topologies; the rigorous way to deal with these expressions and equations can be found in [33]). The link with the notations of the previous sections is $\sqrt{Q}e_k = \sigma^k$.

One of the typical practical problems to be solved is the computation of averages of the form

$$E[\varphi(u(t))]$$

(or more generally of the form $E[\varphi(u(t_1), \ldots, u(t_n))]$), for a given observable φ. For suitable φ's, this allows one to quantify uncertainty (mean and deviations), understand statistical scaling laws (correlation functions, structure function) or compute other average quantities of interest.

The most obvious approach to the (approximate) computation of $E\left[\varphi\left(u\left(t\right)\right)\right]$ is Monte Carlo method: it consists in the direct numerical simulation of the equation, repeated several times for different noise realizations, followed by the arithmetic average of the results. It is difficult to compete with this simple, flexible method. It is however a duty of mathematical research to investigate alternatives. Let us see two of them offered by tools of stochastic analysis. Let us anticipate that we cannot, at present, state that they are better than simple Monte Carlo simulation, so they are presented here for speculation and promotion of further research.

Consider the associated linear equation

$$dz\left(t\right) = Az\left(t\right)dt + \sqrt{Q}dW\left(t\right), \qquad z\left(0\right) = u_0 \qquad (3)$$

(e.g. Stokes equation). Its solution is given by

$$z\left(t\right) = e^{tA}u_0 + \int_0^t e^{(t-s)A}\sqrt{Q}dW\left(s\right).$$

The process $\left(z\left(t\right)\right)_{t\geq 0}$ is Gaussian and there is an explicit formula for its mean function $m_t := E\left[z\left(t\right)\right]$ and covariance function Q_{t_1,t_2}, defined on test functions $h, k \in H$ as

$$\langle Q_{t_1,t_2}h, k\rangle = E\left[\langle z\left(t_1\right) - m_{t_1}, h\rangle\, \langle z\left(t_2\right) - m_{t_2}, k\rangle\right]$$

(the functions m_t and Q_{t_1,t_2} characterize the law of the Gaussian process $\left(z\left(t\right)\right)_{t\geq 0}$) in terms of A and Q:

$$E\left[z\left(t\right)\right] = e^{tA}u_0$$

(namely m_t solves the equation $m_t' = Am_t$) and, for $t_1 \geq t_2 \geq 0$,

$$Q_{t_1,t_2} = \int_0^{t_2} e^{(t_1-s)A}Qe^{(t_2-s)A^*}ds.$$

If we denote by $\mathcal{N}_{\mu,\Sigma}\left(dz\right)$ the law on H of a Gaussian vector of mean μ and covariance Σ, then

$$E\left[\varphi\left(z\left(t\right)\right)\right] = \int_H \varphi\left(z\right)\mathcal{N}_{m_t,Q_{t,t}}\left(dz\right) \qquad (4)$$

and more generally we have a formula for $E\left[\varphi\left(z\left(t_1\right), \ldots, z\left(t_n\right)\right)\right]$. Thus we may compute an approximate value of $E\left[\varphi\left(z\left(t\right)\right)\right]$ by numerical integration of (4) instead of direct numerical simulation of the differential equation (3). Of course, the high or infinite dimension in (4) is again a difficulty, but Monte Carlo is a good method to compute the integral in (4) in high dimensions, and it is much cheaper than Monte Carlo applied directly to the stochastic equation (3).

Thus the question is: can we reduce, approximately, the problem of computation of $E[\varphi(u(t))]$ to the Gaussian case?

Without pretending any originality (the Physics literature certainly contains more advanced ideas), let us mention two approximations based on perturbative ideas. Infinite dimensional stochastic calculus is mature for this kind of investigations and others.

5.1 Perturbative Approach by Girsanov (Cameron-Martin)

This presentation is inspired by the recent work [28], based on ideas of field theory; there, the so called Onsager-Machlup path integrals are used which, from the mathematical side, cannot be made rigorous because a flat measure on paths does not exist; however, the final results of such heuristic theory can be meaningful, case by case, since its finite dimensional approximations could eventually converge. We present here a parallel approach by Girsanov theorem, where Onsager-Machlup integrals are replaced by Wiener averages. See [5] for similar ideas.

Assume Q is invertible. Girsanov theorem, see [19, 22, 33], states that the following identity (sometimes called Cameron-Martin formula) holds

$$E[\varphi(u(t))] = E\left[\varphi(z(t)) e^{\int_0^T \langle Q^{-1/2}B(z(s)),dW(s)\rangle - \frac{1}{2}\int_0^T \|Q^{-1/2}B(z(s))\|^2 ds}\right]$$

under appropriate conditions, for every $T \geq t \geq 0$. The classical one is Novikov condition:

$$E\left[e^{\frac{1}{2}\int_0^T \|Q^{-1/2}B(z(s))\|^2 ds}\right] < \infty$$

but it is too restrictive for fluid dynamics: usually $B(\cdot)$ has quadratic growth and the forth power of a Gaussian variable is not exponentially integrable. A more general result is proved in [10, 11], generalizing the so called Liptser-Shiryaev conditions:

$$P\left(\int_0^T \|Q^{-1/2}B(u(s))\|^2 ds\right) = 1, \qquad P\left(\int_0^T \|Q^{-1/2}B(z(s))\|^2 ds\right) = 1.$$

These conditions are again very severe, because $Q^{-1/2}$ is a de-regularizing operator (usually Q is assumed trace class), B is a differential operator, and thus the operation $\|Q^{-1/2}B(\cdot)\|$ is very demanding, requires $u(t)$ and $z(t)$ to be very regular, but this can be achieved only with additional regularity of Q, hence $Q^{-1/2}$ is more de-regularizing... a satisfactory equilibrium point does not seem to exist, for usual fluid dynamic equations. Nevertheless, the previous scheme could be applied for instance to finite dimensional approximations or other regularizations (e.g. suitable powers of Δ), hence we discuss it.

Assume we have εB in place of B. Girsanov formula

$$E\left[\varphi\left(u\left(t\right)\right)\right] = E\left[\varphi\left(z\left(t\right)\right)e^{\varepsilon\int_0^T\langle Q^{-1/2}B(z(s)),dW(s)\rangle - \frac{\varepsilon^2}{2}\int_0^T\|Q^{-1/2}B(z(s))\|^2 ds}\right]$$

may (formally) be approximated by:

$$= E\left[\varphi\left(z\left(t\right)\right)\right] + \varepsilon E\left[\varphi\left(z\left(t\right)\right)\int_0^T\langle Q^{-1/2}B\left(z\left(s\right)\right),dW\left(s\right)\rangle\right]$$

$$-\frac{\varepsilon^2}{2}E\left[\varphi\left(z\left(t\right)\right)\int_0^T\|Q^{-1/2}B\left(z\left(s\right)\right)\|^2 ds\right]$$

$$+\frac{\varepsilon^2}{2}E\left[\varphi\left(z\left(t\right)\right)\left(\int_0^T\langle Q^{-1/2}B\left(z\left(s\right)\right),dW\left(s\right)\rangle\right)^2\right]+\dots$$

The Gaussian term $E\left[\varphi\left(z\left(t\right)\right)\right]$ has been discussed above. Terms of the form

$$E\left[\varphi\left(z\left(t\right)\right)\int_0^T\psi\left(z\left(s\right)\right)ds\right] \tag{5}$$

can be similarly computed:

$$E\left[\varphi\left(z\left(t\right)\right)\int_0^T\psi\left(z\left(s\right)\right)ds\right] = \int_0^T E\left[\varphi\left(z\left(t\right)\right)\psi\left(z\left(s\right)\right)\right]ds$$

$$= \int_0^T\int_H\int_H\varphi\left(x\right)\psi\left(y\right)\mathcal{N}_{(z(t),z(s))}\left(dx,dy\right)ds.$$

Here we denote by $\mathcal{N}_{(z(t_1),z(t_2))}\left(dx,dy\right)$ the law of $\left(z\left(t_1\right),z\left(t_2\right)\right)$, a Gaussian measure with mean vector (m_{t_1},m_{t_2}) and covariance matrix $\left(Q_{t_i,t_j}\right)_{i,j=1,2}$.

Terms like

$$E\left[\varphi\left(z\left(t\right)\right)\left(\int_0^T\langle\psi\left(z\left(s\right)\right),dW\left(s\right)\rangle\right)^n\right] \tag{6}$$

are more difficult. If $Q^{-1/2}\psi\left(z\right) = DV\left(z\right)$ for a potential V, then

$$\int_0^T\langle\psi\left(z\left(s\right)\right),dW\left(s\right)\rangle = V\left(z\left(T\right)\right) - V\left(z\left(0\right)\right) - \int_0^T\langle Q^{-1/2}\psi\left(z\left(s\right)\right),Az\left(s\right)\rangle ds$$

$$-\frac{1}{2}\int_0^T Tr\left(QD^2V\left(z\left(s\right)\right)\right)ds.$$

This expression can be substituted in (6) and leads again to Gaussian integrals of the form (5) or more complex. This approach is described in [20]; unfortunately, in fluid dynamics it is not common that $Q^{-1/2}B(z) = DV(z)$, since B is more of Hamiltonian than of gradient type.

For $n = 1$, Malliavin calculus offers a brilliant solution to the problem of the computation of (6) (the case $n > 1$, require further research). Using Malliavin "integration by part" calculus, the term (6) can be reduced to

$$\int_0^T E\left[\langle D_s^{\mathcal{M}} \varphi(z(t)), \psi(z(s))\rangle_H\right] ds$$

where $D_s^{\mathcal{M}}$ denotes Malliavin derivative. Using other rules of Malliavin calculus the term $D_s^{\mathcal{M}} \varphi(z(t))$ can be explicitly computed in terms of derivatives of φ and the pair (A, Q) and then the problem reduces again to Gaussian integrations of the form (5). The Malliavin approach and numerical simulations of the whole approximation procedure based on Girsanov formula have been performed for low-dimensional systems (cf. [29]). The result show that additional research in necessary since the range of ε for which the approximation is reasonable and competitive with Monte Carlo simulations is very small ($\varepsilon \sim 0.05$ for very simple systems with quadratic nonlinearities and data having order of magnitude of unity).

5.2   Perturbative Approach by Kolmogorov Equation

Consider again (now we denote explicitly the initial condition u_0) the equation

$$du(t; u_0) = Au(t; u_0) dt + B(u(t; u_0)) dt + \sqrt{Q}dW(t) \qquad u(0; u_0) = u_0$$

and its Gaussian counterpart

$$dz(t; u_0) = Az(t; u_0) dt + \sqrt{Q}dW(t) \qquad z(0; u_0) = u_0.$$

The function

$$U(t, x) = E[\varphi(u(t; x))], \qquad t \geq 0, x \in H$$

satisfies the *Kolmogorov equation*

$$\frac{\partial U}{\partial t} = \frac{1}{2}Tr\left(QD^2U\right) + \langle Ax, DU\rangle + \langle B(x), DU\rangle \qquad U|_{t=0} = \varphi.$$

Under appropriate conditions, this is a rigorous result, also in Hilbert spaces (see [34]). Let us also mention the remarkable work [30] (see also [12] and [17] for variants) where a solution to the Kolmogorov equation associated to 3D stochastic Navier-Stokes equations is provided; in this case $u(t; u_0)$ denotes a Markov selection, as described above.

Denote by P_t the *Ornstein-Uhlenbeck semigroup* defined as

$$P_t \varphi (x) = E\left[\varphi\left(z\left(t; x\right)\right)\right], \qquad t \geq 0, x \in H$$

which gives us the solution $V(t, x) = P_t \varphi(x)$ of the equation

$$\frac{\partial V}{\partial t} = \frac{1}{2} Tr\left(QD^2 V\right) + \langle Ax, DV \rangle \qquad V|_{t=0} = \varphi.$$

It is "explicitly" computable, as we have explained in the previous section, because $z(t, x)$ is Gaussian. Then

$$U(t, x) = P_t \varphi(x) + \int_0^t P_{t-s}\left(\langle B(\cdot), DU(s, \cdot)\rangle\right)(x)\, ds.$$

This equation connects $U(t, x) = E\left[\varphi\left(u\left(t; x\right)\right)\right]$ with P_t applied to different functions; and P_t is explicitly computable.

Under suitable conditions there is hope to prove that the iteration

$$U^0(t, x) = P_t \varphi(x)$$

$$U^{n+1}(t, x) = P_t \varphi(x) + \int_0^t P_{t-s}\left(\langle B(\cdot), DU^n(s, \cdot)\rangle_H\right)(x)\, ds$$

converges exponentially fast to the solution $U(t, x)$. When B is bounded on H, at least for small t the convergence hold, see [34] (for larger t one could introduce an exponential penalization). When B is the transport term of 3D Navier-Stokes equations, this approach fails and a modification is needed, see [30], where the auxiliary Gaussian process $z(t; u_0)$ is replaced by another one $\tilde{z}(t; u_0)$, no more Gaussian; the computability of the modified auxiliary process $\tilde{z}$ is an issue for future research, that we cannot discuss here.

Then (in the case when the Gaussian process $z(t; u_0)$ may be used, e.g. for finite dimensional or regularized approximations of the 3D Navier-Stokes equations) a few iterations provide a computable Gaussian approximation. Indeed,

$$U^0(t, x) = P_t \varphi(x) = E\left[\varphi\left(z\left(t; x\right)\right)\right]$$

$$U^1(t, x) = P_t \varphi(x) + \int_0^t P_{t-s}\left(\langle B(\cdot), DU^0(s, \cdot)\rangle_H\right)(x)\, ds$$

$$= E\left[\varphi\left(z\left(t; x\right)\right)\right] + \int_0^t E\left[\langle B\left(z\left(t-s; x\right)\right), DP_s \varphi\left(z\left(t-s; x\right)\right)\rangle_H\right]\, ds$$

and there is an "explicit formula" for $DP_s\varphi$, see [34]. And so on. Similarly to the expansion of Girsanov formula, we have an approximation of

$$U(t,x) = E\left[\varphi\left(u\left(t;x\right)\right)\right]$$

by means of Gaussian integrals.

We have discussed so far the perturbation of Kolmogorov equations but similar ideas could be developed for the Fokker-Planck equations, which describe the time evolution of probability law of solutions; see for instance [6] as a foundational work in infinite dimensions.

To conclude this illustration of ideas, let us remark that almost all aspects of this research have still to be done: rigorous error estimates, practical realizations, estimates of computations cost, application to fluid dynamic models.

References

1. Albeverio, S., Ferrario, B.: Some methods of infinite dimensional analysis in hydrodynamics: an introduction. In: SPDE in Hydrodynamic: Recent Progress and Prospects. Lecture Notes in Mathematics, vol. 1942, pp. 1–50. Springer, Berlin (2008)
2. Barbato, D., Flandoli, F., Morandin, F.: Uniqueness for a stochastic inviscid dyadic model. Proc. Am. Math. Soc. **138**(7), 2607–2617 (2010)
3. Barbato, D., Flandoli, F., Morandin, F.: Anomalous dissipation in a stochastic inviscid dyadic model. Ann. Appl. Probab. **21**(6), 2424–2446 (2011)
4. Barbato, D., Morandin, F.: Stochastic inviscid shell models: well-posedness and anomalous dissipation. Nonlinearity **26**(7), 1919–1943 (2013)
5. Bertini, L., Jona-Lasinio, G., Parrinello, C.: Stochastic quantization, stochastic calculus and path integrals: selected topics. Progr. Theor. Phys. Suppl. No. **111**, 83–113 (1993)
6. Bogachev, V.I., Da Prato, G., Röckner, M.: Existence and uniqueness of solutions for Fokker-Planck equations in Hilbert spaces. J. Evol. Equ. **10**, 487–509 (2010)
7. Chow, P.-L.: Stochastic Partial Differential Equations. Applied Mathematics and Nonlinear Science. Chapman & Hall/CRC, Boca Raton (2007)
8. Debussche, A., Odasso, C.: Markov solutions for the 3D stochastic Navier-Stokes equations with state dependent noise. J. Evol. Equ. **6**(2), 305–324 (2006)
9. Fefferman, C.L.: Existence and Smoothness of the Navier-Stokes Equations. The Millennium Prize Problems, pp. 57–67. Clay Mathematics Institute, Cambridge (2006)
10. Ferrario, B.: Absolute continuity of laws for semilinear stochastic equations with additive noise. Commun. Stoch. Anal. **2**(2), 209–227 (2008)
11. Ferrario, B.: A note on a result of Liptser-Shiryaev. Stoch. Anal. Appl. **30**(6), 1019–1040 (2012)
12. Flandoli, F.: On the method of Da Prato and Debussche for the 3D stochastic Navier Stokes equations. J. Evol. Equ. **6**(2), 269–286 (2006)
13. Flandoli, F.: An introduction to 3D stochastic fluid dynamics. In: SPDE in Hydrodynamic: Recent Progress and Prospects. Lecture Notes in Mathematics, vol. 1942, pp. 51–150. Springer, Berlin (2008)
14. Flandoli, F.: Random perturbation of PDEs and fluid dynamic models. In: Saint Flour Summer School Lectures 2010. Lecture Notes in Mathematics, vol. 2015. Springer, Berlin (2011)
15. Flandoli, F., Gatarek, D.: Martingale and stationary solutions for stochastic Navier-Stokes equations. Probab. Theory Relat. Fields **102**(3), 367–391 (1995)

16. Flandoli, F., Gubinelli, M., Priola, E.: Well posedness of the transport equation by stochastic perturbation. Invent. Math. **180**, 1–53 (2010)
17. Flandoli, F., Romito, M.: Regularity of transition semigroups associated to a 3D stochastic Navier-Stokes equation. In: Baxendale, P.H., Lototsky, S.V. (eds) Stochastic Differential Equations: Theory and Applications. Interdisciplinary Mathematical Sciences, vol. 2, pp. 263–280. World Scientific Publishing, Hackensack (2007)
18. Flandoli, F., Romito, M.: Markov selections for the 3D stochastic Navier-Stokes equations. Probab. Theory Relat. Fields **140**(3–4), 407–458 (2008)
19. Goldys, B., Maslowski, B.: Lower estimates of transition densities and bounds on exponential ergodicity for stochastic PDE's. Ann. Probab. **34**(4), 1451–1496 (2006)
20. Jona-Lasinio, G., Sénéor, R.: Study of stochastic differential equations by constructive methods. I. J. Stat. Phys. **83**(5–6), 1109–1148 (1996)
21. Kotelenez, P.: Stochastic Ordinary and Stochastic Partial Differential Equations: Transition from Microscopic to Macroscopic Equations. Stochastic Modelling and Applied Probability, vol. 58. Springer, New York (2008)
22. Kozlov, S.M.: Some questions on stochastic equations with partial derivatives. Trudy Sem. Petrovsk. **4**, 147–172 (1978)
23. Krylov, N.V., Röckner, M.: Strong solutions of stochastic equations with singular time dependent drift. Probab. Theory Relat. Fields **131**, 154–196 (2005)
24. Kuksin, S.B., Shirikyan, A.: Mathematics of Two-Dimensional Turbulence. Cambridge University Press, Cambridge (2012)
25. Kupiainen, A.: Lessons for turbulence. Geom. Funct. Anal. Spec Vol GAFA-2000, 316–333 (2000)
26. Lions, P.L.: Mathematical Topics in Fluid Mechanics. Incompressible Models, vol. 1, Oxford University Press, New York (1996)
27. Mikulevicius, R., Rozovskii, B.L.: Global L^2-solutions of stochastic Navier-Stokes equations. Ann. Probab. **33**(1), 137–176 (2005)
28. Navarra, A., Tribbia, J., Conti, G.: The path integral formulation of climate dynamics. CMCC Research Paper No. 130 (2012)
29. Pratelli, B.: Master thesis, Pisa (2013)
30. Da Prato, G., Debussche, A.: Ergodicity for the 3D stochastic Navier-Stokes equations. J. Math. Pures Appl. (9) **82**(8), 877–947 (2003)
31. Da Prato, G., Flandoli, F.: Pathwise uniqueness for a class of SDE in Hilbert spaces and applications. J. Funct. Anal. **259**(1), 243–267 (2010)
32. Da Prato, G., Flandoli, F., Priola, E., Röckner, M.: Strong uniqueness for stochastic evolution equations in Hilbert spaces perturbed by a bounded measurable drift. Ann. Probab. **41**, 3306–3344 (2013)
33. Da Prato, G., Zabczyk, J.: Stochastic Equations in Infinite Dimensions. Cambridge University Press, Cambridge (1992)
34. Da Prato, G., Zabczyk, J.: Second Order Partial Differential Equations in Hilbert Spaces. London Mathematical Society Lecture Note Series, vol. 293. Cambridge University Press, Cambridge (2002)
35. Romito, R.: Analysis of equilibrium states of Markov solutions to the 3D Navier-Stokes equations driven by additive noise. J. Stat. Phys. **131**(3), 415–444 (2008)
36. Veretennikov, Y.A.: On strong solution and explicit formulas for solutions of stochastic integral equations. Math. USSR Sb. **39**, 387–403 (1981)
37. Visik, M.I., Fursikov, A.V.: Mathematical Problems of Statistical Hydromechanics. Akademische Verlagsgesellschaft Geest & Portig K.-G., Leipzig (1986)

Remarks on Long Time Versus Steady State Optimal Control

Alessio Porretta and Enrique Zuazua

Abstract Control problems play a key role in many fields of Engineering, Economics and Sciences. This applies, in particular, to climate sciences where, often times, relevant problems are formulated in long time scales. The problem of the possible asymptotic simplification (as time tends to infinity) then emerges naturally. More precisely, assuming, for instance, that the free dynamics under consideration stabilizes towards a steady state solution, the following question arises: Do time averages of optimal controls and trajectories converge to the steady optimal controls and states as the time-horizon tends to infinity?

This question is very closely related to the so-called *turnpike* property stating that, often times, the optimal trajectory joining two points that are far apart, consists in, departing from the point of origin, rapidly getting close to the steady-state (the turnpike) to stay there most of the time, to quit it only very close to the final destination and time.

In this paper we focus on the semilinear heat equation. We prove some partial results and enumerate a number of interesting topics of future research, indicating also some connections with shape design and inverse problems theory.

Keywords Semilinear heat equations • Optimal control problems • Long time behavior • Steady states • Controllability • Observability • Turnpike property

AMS subject classification: 49J20, 49K20, 93C20, 49N05

A. Porretta (✉)
Dipartimento di Matematica, Università di Roma Tor Vergata, Via della ricerca scientifica 1, 00133 Roma, Italy
e-mail: porretta@mat.uniroma2.it

E. Zuazua
Departamento de Matemáticas, Universidad Autónoma de Madrid, 28049 Madrid, Spain
e-mail: enrique.zuazua@uam.es

© Springer International Publishing Switzerland 2016
F. Ancona et al. (eds.), *Mathematical Paradigms of Climate Science*, Springer
INdAM Series 15, DOI 10.1007/978-3-319-39092-5_5

1 Introduction

In this paper, we address the question of the limiting behavior of optimal control problems as the time-horizon tends to infinity for semilinear heat equations. Although the question makes sense and is relevant for a much wider class of problems, we focus on this particular case to simplify the presentation, and to underline some of the main difficulties one encounters when addressing these problems and the fundamental tools needed in their analysis.

The motivation to consider this kind of problems is clear in many contexts but in particular in climate sciences where problems are naturally formulated in long time intervals. This is for instance the case in paleoclimatology (study of past climates) (see, for instance, [14]) where the problem of the inversion of past climates is addressed.

Note however that the models arising in climate sciences are extremely complex. Thus, rigorously speaking, although the topic addressed here can be of relevance in that field, the techniques we develop cannot be directly applied and will require significant further developments.

Sustainable economic development is another area in which these issues arises, playing a central role (see [7]).

Most often, the existing Partial Differential Equations (PDE) Control Theory, based on optimization and minimization of cost functionals, and the characterization of optimal controls through the corresponding optimality systems and adjoint methods, does not distinguish between short and long time horizons.

Here we are specifically interested in long-time horizon control problems and the possibility that optimal trajectories and controls simplify towards those of the corresponding steady state model.

In practice, in long time-horizons, the effective computation of the control can be very expensive since it requires iterative methods to solve the coupled optimality system combining the forward controlled state equation and the backward adjoint one.

It is then natural to look for some shortcuts. This makes sense, in particular, when, as it occurs often times in applications, the free dynamics associated to the state equation presents some property of asymptotic simplification: convergence towards a steady state solution, stabilization around a periodic trajectory or a self-similar solution, etc. When that occurs it is natural to investigate whether the optimal control and trajectories converge towards the corresponding simplified optimal control and states.

In other words, the question we are discussing consists in analyzing whether the processes of long time asymptotics and control commute.

This problem, as pointed out in [15], is related to the so-called *turnpike property*, mainly motivated by economic theories (see [22] and references therein). We also refer to the more recent paper on time-discrete finite-dimensional systems [10] and [9], the seminal continuous time paper [2] and the more recent and systematic one [20].

The question of whether the control process commutes with some qualitative aspect of PDE models has been analyzed in other contexts too. For instance, it is well known that the question is very subtle when dealing with numerical approximation methods. More precisely, convergent numerical algorithms for the free dynamics do not necessarily lead to convergent numerical methods for control problems, especially, when one is dealing with the more demanding problem of controllability (see [23]). This is so, in particular, when the numerical scheme is not stable enough to avoid the emergence of spurious numerical high frequency solutions. A certain amount of dissipativity of the numerical schemes is required and, as we shall see, the same can be said when dealing with the long-time horizon control problems.

The issue of long time versus steady state control is also relevant in shape design. In particular, in the field of aeronautics, most designs are computed based on steady state models and, although it is assumed or understood that these steady optimal shapes are close to the optimal time-evolving ones, there are not results justifying such a fact rigorously, especially for the relevant models in fluid mechanics such as Navier-Stokes or Euler equations (see [12]). Similar questions also arise in the context of inverse problems (see [11], section 9.4 and [8]).

In our earlier paper [15] we addressed the problem of long time horizon versus steady state control in the linear setting. There we analyzed in detail both the finite-dimensional case, and the paradigmatic PDE models, namely, the heat and the wave equations, and proved that, under suitable controllability conditions (see [24] for a general presentation of the theory of controllability for PDE), optimal controls and controlled trajectories converge, as the time horizon tends to infinity, towards the stationary optimal controls and states with an exponential rate induced by the stabilizing Riccati feedback operator.

The analysis in [15], however, is of purely linear nature, based on the properties of the optimality system characterizing the optimal controls and states through the coupling with the adjoint system.

But the problem makes sense in the nonlinear context too.

In this paper we briefly discuss the issue for the semilinear heat equation. We first present the main tools developed in [15] in the linear case, in order to later employ them to get results of local nature for the semilinear heat equation. We then consider the simpler problem of time-independent controls showing how simpler and more classical Γ-convergence arguments allow to handle it. This late result, although much simpler to be achieved, is also relevant from the point of view of applications, where the applied controls can be time-independent as well.

We close the paper formulating a number of open problems and directions of future research.

2 Preliminaries on the Linear Heat Equation

Let us briefly recall the main results obtained in [15] in the specific case of the following controlled heat equation.

Let $\Omega \subset \mathbb{R}^N$ be a bounded domain and consider the heat equation with Dirichlet boundary conditions and an applied control:

$$\begin{cases} y_t - \Delta y = u\chi_\omega & \text{in } (0, T) \times \Omega \\ y = 0 & \text{on } (0, T) \times \partial\Omega \\ y(0) = y_0 \in L^2(\Omega) . \end{cases} \tag{1}$$

Then, consider the following associated control problem

$$\min J^T(u) = \frac{1}{2} \int_0^T \left[|u(t)|^2_{L^2(\omega)} + |y(t) - z|^2_{L^2(\omega_0)} \right] dt \tag{2}$$

where $u \in L^2(0, T; L^2(\omega))$ and y solves (1), and $z \in L^2(\omega_0)$ is a given observation. Here ω and ω_0 are two open subsets of Ω and χ_ω stands for the characteristic function of the set ω where the control is being applied, while ω_0 denotes the subdomain where the tracking term of the cost functional is active.

We also consider the stationary version of the state equation:

$$\begin{cases} -\Delta y = u\chi_\omega & \text{in } \Omega \\ y = 0 & \text{on } \Omega, \end{cases} \tag{3}$$

and the corresponding problem of minimizing the functional

$$\min J(u) = \frac{1}{2} \left[|u|^2_{L^2(\omega)} + |y - z|^2_{L^2(\omega_0)} \right]. \tag{4}$$

Let us now consider the control problems (2) and (4) and the corresponding optimal solutions (u^T, y^T) and $(\bar{u}, \bar{y})$, respectively. Then, according to the results in [15], there exists $\mu > 0$ such that

$$\|y^T(t) - \bar{y}\|_{L^2(\Omega)} + \|u^T(t) - \bar{u}\|_{L^2(\Omega)} \le K(e^{-\mu t} + e^{-\mu(T-t)}) \tag{5}$$

for every $t \in [0, T]$. Let us now sketch the main steps of the proof that will give us a precise idea of the constants K, μ involved in this estimate.

The optimality systems for the time evolution and steady state problems read as follows, respectively:

$$\begin{cases} y_t^T - \Delta y^T = -q^T \chi_\omega & \text{in } \Omega \times (0, T) \\ y^T = 0 & \text{on } \partial\Omega \times (0, T) \\ y^T(0) = y_0 \\ -q_t^T - \Delta q^T = (y^T - z)\chi_{\omega_0} & \text{in } \Omega \times (0, T) \\ q^T = 0 & \text{on } \partial\Omega \times (0, T) \\ q^T(T) = 0, \end{cases} \tag{6}$$

and

$$\begin{cases} -\Delta \bar{y} = -\bar{q}\chi_\omega & \text{in } \Omega \\ \bar{y} = 0 & \text{on } \partial\Omega \\ -\Delta \bar{q} = (\bar{y} - z)\chi_{\omega_0} & \text{in } \Omega \\ \bar{q} = 0 & \text{on } \partial\Omega. \end{cases} \tag{7}$$

In the reference case where $z = 0$, we define a linear bounded operator in $L^2(\Omega)$ as

$$\mathscr{E}(T)y_0 := q^T(0) \ .$$

It turns out that $\mathscr{E}(t)$ is a positive operator which is increasing and uniformly bounded with respect to t, and we have

$$\|\mathscr{E}(t) - \hat{E}\|_{\mathscr{L}(L^2(\Omega),L^2(\Omega))} \leq Ce^{-\mu t}, \tag{8}$$

for some $C > 0$ and $\mu > 0$, where $\hat{E}$ is the corresponding operator for the infinite horizon control problem with $z = 0$. Namely,

$$\hat{E}y_0 := \hat{q}(0)$$

where, in this case, the pair $(\hat{y}, \hat{q})$ solves the optimality system in infinite time associated with $z = 0$:

$$\begin{cases} \hat{y}_t - \Delta \hat{y} = -\hat{p}\chi_\omega & \text{in } \Omega \times (0, \infty) \\ \hat{y} = 0 & \text{on } \partial\Omega \times (0, \infty) \\ \hat{y}(0) = y_0 \\ -\hat{q}_t - \Delta \hat{q} = \hat{y}\chi_{\omega_0} & \text{in } \Omega \times (0, \infty) \\ \hat{q} = 0 & \text{on } \partial\Omega \times (0, \infty) \\ \|\hat{q}(t)\|_{L^2(\Omega)} \to 0 & \text{as } t \to \infty. \end{cases} \tag{9}$$

Notice that, by time invariance, we have $\hat{q}(t) = \hat{E}\hat{y}(t)$, and the first equation in (9) defines an operator $M := -\Delta + \hat{E} \chi_\omega$ which is exponentially stable, providing the rate μ which appears in (8), as well as in (5).

Once the operators $\mathscr{E}(t)$ and $\hat{E}$ are defined as above in terms of the reference problem where $z = 0$, the adjoint state of the general system (6) can be represented by the affine feedback law

$$q^T(t) - \bar{q} = \mathscr{E}(T - t)(y^T(t) - \bar{y}) + h^T(t) \tag{10}$$

where h^T solves

$$
\begin{cases}
-h_t^T + (-\Delta + \mathscr{E}(T-t)\chi_\omega)h^T = 0 & \text{in } \Omega \times (0,T) \\
h^T = 0 & \text{on } \partial\Omega \times (0,T) \\
h^T(T) = -\bar{q}.
\end{cases}
\tag{11}
$$

In some sense, h^T is a kind of corrector taking care of the final cost at time T; and the equation satisfied by h^T can be deduced from equality (10) using the Riccati equation satisfied by $\mathscr{E}(t)$ and the optimality system (6). Alternatively, instead of using the Riccati equation, one can define h^T as a solution of (11) and verify a posteriori that equality (10) holds in a weak sense

$$
\int_\Omega (q^T(t) - \bar{q})\varphi\, dx = \int_\Omega (y^T(t) - \bar{y})[\mathscr{E}(T-t)\varphi]\, dx + \int_\Omega h^T(t)\,\varphi\, dx
$$

using properly the definition of $\mathscr{E}(t)$.

The corrector h^T can be estimated from (8) and the exponential stability of $-\Delta + \hat{E}\chi_\omega$; in fact, one can prove that

$$
||h^T(t)||_{L^2(\Omega)} \le C^* ||\bar{q}||_{L^2(\Omega)} e^{-\mu(T-t)}.
\tag{12}
$$

Once this structure is observed, the system (6) can be uncoupled by writing that the optimal trajectory y^T solves

$$
y_t^T - \Delta y^T = -q^T\chi_\omega = -\bar{q}\chi_\omega - \mathscr{E}(T-t)\chi_\omega(y^T(t) - \bar{y}) - h^T\chi_\omega
$$

which implies

$$
(y^T(t) - \bar{y})_t - \Delta(y^T(t) - \bar{y}) = -\mathscr{E}(T-t)\chi_\omega(y^T(t) - \bar{y}) - h^T\chi_\omega\,.
$$

The exponential proximity property (5) is now a straightforward consequence of (8), (12) and the decay of the stabilized dynamics. In fact, we have

$$
(y^T(t) - \bar{y})_t + [-\Delta + \hat{E}\chi_\omega](y^T(t) - \bar{y}) = (\hat{E} - \mathscr{E}(T-t))\chi_\omega(y^T(t) - \bar{y}) - h^T\chi_\omega
$$

which implies estimate (5) in the more precise form

$$
||y^T(t) - \bar{y}||_{L^2(\Omega)} + ||u^T(t) - \bar{u}||_{L^2(\Omega)} \le \tilde{K}(||y_0 - \bar{y}||_{L^2(\Omega)} e^{-\mu t} + ||\bar{q}||e^{-\mu(T-t)})\,,
\tag{13}
$$

for every $t \in [0, T]$. Here the constant $\tilde{K}$ is independent of the choice of the initial data and of the target z.

Remark 1 Let us stress that (13) would take a more symmetric form if the adjoint state p^T had a different prescribed data at time $t = T$. This is the case if we consider a cost functional with an additional final pay-off such as

$$J^T(u) = \frac{1}{2} \int_0^T \left[|u(t)|^2_{L^2(\omega)} + |y(t) - z|^2_{L^2(\omega_0)} \right] dt + q_0 \cdot y(T)$$

for some $q_0 \in L^2(\Omega)$. In this case, the adjoint state q^T must satisfy $q^T(T) = q_0$; the above proof applies without changes except that now the corrector term h^T will take a different final condition (equal to $q_0 - \bar{q}$) and the estimate (13) would become

$$\|y^T(t) - \bar{y}\|_{L^2(\Omega)} + \|u^T(t) - \bar{u}\|_{L^2(\Omega)} \leq \tilde{K}(\|y_0 - \bar{y}\|_{L^2(\Omega)} e^{-\mu t} + \|q_0 - \bar{q}\| e^{-\mu(T-t)}),$$

for every $t \in [0, T]$.

The above remark points out that the exponential turnpike property is somehow symmetric with respect to what happens at $t = 0$ and $t = T$, see also a more general discussion of this fact in [20] for the finite-d case.

3 Local Results for the Semilinear Heat Equation

This section is divided in two parts. In the first one we formulate the problem under consideration and prove a turnpike property for the system of optimality under the condition that the initial and final states are close enough to the stationary primal and dual state, respectively. In the second subsection, we show, as an example, that this result applies at least in the case that the target and the initial datum are small enough. We stress, however, that this solution of the system of optimality is not guaranteed to be a minimizer for the time-dependent optimal control problem, unless one proves that the functional under consideration is (locally) convex. A similar strategy was used in [4] to analyze the optima of a nonlinear control problem arising in mean field games theory; unfortunately, for the semilinear heat equation considered below, the convexity of the functional is not clear and the equivalence between minima and solutions of the optimality system requires further investigation.

3.1　The System of Optimality

It is natural to consider the same issues of the previous section for the following semilinear heat equation:

$$\begin{cases} y_t - \Delta y + f(y) = u\chi_\omega & \text{in } \Omega \times (0, T) \\ y = 0 & \text{on } \partial\Omega \times (0, T) \\ y(0) = y_0 \in L^2(\Omega) , \end{cases} \tag{14}$$

f being a C^1 nondecreasing function.

The semilinear problem (14) is well-posed. More precisely, given $y_0 \in L^2(\Omega)$ and $u \in L^2(\omega \times (0, T))$, there exists a unique solution

$$y \in C([0, T]; L^2(\Omega)) \cap L^2(0, T; H_0^1(\Omega)).$$

We consider the optimal control problem:

$$\min \{J^T(u) := \frac{1}{2} \int_0^T |y(t) - z|^2_{L^2(\omega_0)} dt + \frac{1}{2} \int_0^T |u|^2_{L^2(\omega)} dt + q_0 \cdot y(T)\} , \tag{15}$$

where $z \in L^2(\omega_0)$ and $q_0 \in L^2(\Omega)$. In the stationary version the state equation is

$$\begin{cases} -\Delta y + f(y) = u\chi_\omega & \text{in } \Omega \\ y = 0 & \text{on } \partial\Omega , \end{cases} \tag{16}$$

together with the corresponding functional

$$\min \{J(u) := \frac{1}{2} \left[|y - z|^2_{L^2(\omega_0)} + |u|^2_{L^2(\omega)} \right] \} . \tag{17}$$

In both cases it is easy to see that the optima are achieved and we can easily write the corresponding optimality systems. They read as

$$\begin{cases} y_t^T - \Delta y^T + f(y^T) = -q^T \chi_\omega & \text{in } \Omega \times (0, T) \\ y^T = 0 & \text{on } \partial\Omega \times (0, T) \\ y^T(0) = y_0 \\ -q_t^T - \Delta q^T + f'(y^T)q^T = (y^T - z)\chi_{\omega_0} & \text{in } \Omega \times (0, T) \\ q^T = 0 & \text{on } \partial\Omega \times (0, T) \\ q^T(T) = q_0, \end{cases} \tag{18}$$

and

$$
\begin{cases}
-\Delta \bar{y} + f(\bar{y}) = -\bar{q}\chi_\omega & \text{in } \Omega \\
\bar{y} = 0 & \text{on } \partial\Omega \\
-\Delta \bar{q} + f'(\bar{y})\bar{q} = (\bar{y} - z)\chi_{\omega_0} & \text{in } \Omega \\
\bar{q} = 0 & \text{on } \partial\Omega.
\end{cases}
\tag{19}
$$

But, due to the nonlinearity of the problems under consideration, the methods of the previous section cannot be applied directly. Hence we develop a local analysis around a given steady state optimal control.

Thus, let $(\bar{y}, \bar{u})$ be an optimal pair for the steady-state problem and introduce the change of variables:

$$
\eta = y^T - \bar{y}; \quad \varphi = q^T - \bar{q}.
$$

Then, (η, φ) satisfy:

$$
\begin{cases}
\eta_t - \Delta\eta + F(\eta) = -\varphi\chi_\omega & \text{in } (0, T) \times \Omega \\
\eta = 0 & \text{on } (0, T) \times \partial\Omega \\
\eta(0) = \eta_0 & \text{in } \Omega \\
-\varphi_t - \Delta\varphi + \Phi(\eta, \varphi) = \eta\chi_{\omega_0} & \text{in } (0, T) \times \Omega \\
\bar{q} = 0 & \text{on } (0, T) \times \partial\Omega \\
\varphi(T) = \varphi_0 & \text{in } \Omega
\end{cases}
\tag{20}
$$

where

$$
\eta_0 = y_0 - \bar{y}, \qquad \varphi_0 = q_0 - \bar{q}
$$

and

$$
F(\bar{y}, \eta) = f(\bar{y} + \eta) - f(\bar{y}),
$$
$$
\Phi(\eta, \varphi) = f'(\bar{y} + \eta)(\bar{q} + \varphi) - f'(\bar{y})\bar{q}
$$

Our aim is to build a pair (u^T, y^T) fulfilling the turnpike property, i.e. such that $(u^T, y^T) \sim (\bar{u}, \bar{y})$ in the sense of the previous section. In the (η, φ) variables, this is equivalent to finding $(\eta, \varphi) \sim (0, 0)$. It is therefore natural to look at the linearization of the above functions F, Φ near $(\eta = 0, \varphi = 0)$ and at the corresponding linearised optimality system:

$$\begin{cases} \eta_t - \Delta\eta + f'(\bar{y})\eta = -\varphi\chi_\omega & \text{in } (0,T) \times \Omega \\ \eta = 0 & \text{on } (0,T) \times \partial\Omega \\ \eta(0) = \eta_0 & \text{in } \Omega \\ -\varphi_t - \Delta\varphi + f'(\bar{y})\varphi = \eta\chi_{\omega_0} - f''(\bar{y})\bar{q}\,\eta & \text{in } (0,T) \times \Omega \\ \bar{q} = 0 & \text{on } (0,T) \times \partial\Omega \\ \varphi(T) = 0 & \text{in } \Omega. \end{cases} \tag{21}$$

It defines a bounded (linear) feedback in $L^2(\Omega)$ as

$$\mathscr{E}(T)\eta_0 = \varphi(0).$$

Let us assume that for some $C > 0$ and $\mu > 0$, we have

$$\begin{aligned} &\|\mathscr{E}(t) - \hat{E}\|_{\mathscr{L}(L^2(\Omega),L^2(\Omega))} \le Ce^{-\mu t}, \\ &\|e^{-tM}\|_{\mathscr{L}(L^2(\Omega),L^2(\Omega))} \le e^{-\mu t}, \quad M := -\Delta + f'(\bar{y}) + \hat{E}\chi_\omega \end{aligned} \tag{22}$$

where $\hat{E}$ is the corresponding operator for the problem in $(0,\infty)$.

Using this exponential stability property of the linearized system, we will be able to prove the following statement. For simplicity, we assume that $f \in C^3$ and the dimension $n \le 3$.

Theorem 1 *Assume that system (21) satisfies the exponential turnpike property. Then, there exists some $\varepsilon > 0$ such that for every y_0, q_0 with*

$$\|y_0 - \bar{y}\|_{L^\infty(\Omega)} + \|q_0 - \bar{q}\|_{L^\infty(\Omega)} \le \varepsilon,$$

there exists a solution of the optimality system

$$\begin{cases} y_t^T - \Delta y^T + f(y^T) = -q^T\chi_\omega & \text{in } \Omega \times (0,T) \\ y^T = 0 & \text{on } \partial\Omega \times (0,T) \\ y^T(0) = y_0 & \\ -q_t^T - \Delta q^T + f'(y^T)q^T = (y^T - z)\chi_{\omega_0} & \text{in } \Omega \times (0,T) \\ q^T = 0 & \text{on } \partial\Omega \times (0,T) \\ q^T(T) = q_0, \end{cases} \tag{23}$$

which satisfies

$$\|y^T(t) - \bar{y}\|_{L^\infty(\Omega)} + \|q^T(t) - \bar{q}\|_{L^\infty(\Omega)} \le K(e^{-\mu t} + e^{-\mu(T-t)}), \forall 0 < t < T. \tag{24}$$

As mentioned above, the turnpike property is established for solutions of the optimality system. Thus, the result does not have the nature we expect, in other words, it does not really apply to the minimizers of the functional under consideration. Dealing with minimizers requires further analysis as the one developed in [20] in the finite-dimensional case.

As we shall see in the following section, this theorem applies at least when the target z is small enough, in which case the steady state problem has a unique minimum which is also small and system (21) satisfies the exponential turnpike property. In this special case, one could expect the solution of the parabolic optimality system to be also unique and to coincide with the optimal state and control.

Proof We look at (20) as a perturbation of the linear system (21) and we aim at finding a solution through a fixed point argument. Namely, for $\hat{\eta}, \hat{\varphi}$ given, we set

$$R_1(\hat{\eta}) := - \left\{ f(\bar{y} + \hat{\eta}) - f(\bar{y}) - f'(\bar{y})\hat{\eta} \right\} ,$$

$$R_2(\hat{\eta}, \hat{\varphi}) := -\bar{q} \left\{ f'(\bar{y} + \hat{\eta}) - f'(\bar{y}) - f''(\bar{y})\hat{\eta} \right\} + [f'(\bar{y}) - f'(\bar{y} + \hat{\eta})]\hat{\varphi}$$

and we define the operator

$$(\eta, \varphi) := \mathscr{K}(\hat{\eta}, \hat{\varphi})$$

where (η, φ) solve

$$\begin{cases} \eta_t - \Delta\eta + f'(\bar{y})\eta = -\varphi\chi_\omega + R_1(\hat{\eta}) & \text{in } (0,T) \times \Omega \\ \eta = 0 & \text{on } (0,T) \times \partial\Omega \\ \eta(0) = \eta_0 & \text{in } \Omega \\ -\varphi_t - \Delta\varphi + f'(\bar{y})\varphi = \eta\chi_{\omega_0} - f''(\bar{y})\bar{q}\,\eta + R_2(\hat{\eta}, \hat{\varphi}) & \text{in } (0,T) \times \Omega \\ \varphi = 0 & \text{on } (0,T) \times \partial\Omega \\ \varphi(T) = \varphi_0 & \text{in } \Omega. \end{cases} \tag{25}$$

with $\eta_0 = y_0 - \bar{y}$, $\varphi_0 = q_0 - \bar{q}$.

Notice that a fixed point would solve the system (20), hence $y^T = \bar{y} + \eta$ and $q^T = \bar{q} + \varphi$ provide a solution of (23).

Assume that $(\hat{\eta}, \hat{\varphi}) \in X$, where

$$X = \left\{ (\eta, \varphi) : \|\eta(t)\|_\infty + \|\varphi(t)\|_\infty \leq M(e^{-\mu t} + e^{-\mu(T-t)}) \quad \forall t \in [0,T] \right\}$$

for some $M \leq 1$. Notice that

$$\begin{aligned} \|R_1(\hat{\eta})(t)\|_2 &\leq C\|R_1(\hat{\eta})(t)\|_\infty \leq c_0 M^2 (e^{-2\mu t} + e^{-2\mu(T-t)}) , \\ \|R_2(\hat{\eta}, \hat{\varphi})(t)\|_2 &\leq C\|R_2(\hat{\eta}, \hat{\varphi})(t)\|_\infty \leq c_1 M^2 (e^{-2\mu t} + e^{-2\mu(T-t)}) \end{aligned} \tag{26}$$

where c_0, c_1 depend on $\|f\|_{C^3[-\|\bar{y}\|_\infty - 1, \|\bar{y}\|_\infty + 1]}$ and on $\|\bar{q}\|_\infty$.

We first remark the following: by defining

$$h^T := \varphi - \mathscr{E}(T - t)\eta$$

then h^T solves the problem

$$\begin{cases} -h_t^T - \Delta h^T + f'(\bar{y})h^T + \mathscr{E}(T - t)\chi_\omega \, h^T = \mathscr{E}(T - t)R_1(\hat{\eta}) + R_2(\hat{\eta}, \hat{\varphi}) & \text{in } (0, T) \times \Omega \\ h^T = 0 & \text{on } (0, T) \times \partial\Omega \\ h^T(T) = \varphi_0 & \text{in } \Omega. \end{cases}$$

We estimate h^T as in Sect. 2. In particular, we have

$$h^T(t) = e^{-M(T-t)}(\varphi_0) + \int_t^T e^{M(t-s)}[(\hat{E} - \mathscr{E}(T - s))\chi_\omega h^T(s)ds$$

$$+ \int_t^T e^{M(t-s)}[\mathscr{E}(T - s)R_1(\hat{\eta}) + R_2(\hat{\eta}, \hat{\varphi})]ds$$

where $M := -\Delta + f'(\bar{y}) + \hat{E}\chi_\omega$. Since M is exponentially stable (with rate μ) and (22) holds, and by means of (26), we get

$$\|h^T(t)\|_2 \leq e^{-\mu(T-t)}\|\varphi_0\|_2 + \int_t^T e^{\mu(t-s)}e^{-\mu(T-s)}\|h^T(s)\|_2 ds$$

$$+ cM^2 \int_t^T e^{\mu(t-s)}[e^{-2\mu(T-s)} + e^{-2\mu s}]ds \,,$$

hence

$$\|h^T(t)\|_2 \leq e^{-\mu(T-t)}[\|\varphi_0\|_2 + cM^2] + cM^2 e^{-2\mu t} + e^{-\mu(T-t)}\int_t^T \|h^T(s)\|_2 ds \,.$$

This implies that

$$\|h^T(t)\|_2 \leq c\,[\|\varphi_0\|_2 + cM^2]e^{-\mu(T-t)} + cM^2 e^{-2\mu t} \,.$$

Coming back to system (25), the first equation now reads as

$$\eta_t - \Delta\eta + f'(\bar{y})\eta + \hat{E}\chi_\omega\eta = -(\mathscr{E}(T - t) - \hat{E})\chi_\omega\eta + R_1(\hat{\eta}) - h^T\chi_\omega$$

so that

$$\eta(t) = e^{-Mt}\eta_0 - \int_0^t e^{-M(t-s)}(\mathscr{E}(T-s) - \hat{E})\chi_\omega \eta(s)ds$$

$$+ \int_0^t e^{-M(t-s)}[R_1(\hat{\eta})(s) - h^T(s)\chi_\omega]ds .$$

On account of (22), (26) and the estimate on h we get

$$\|\eta(t)\|_2 \le e^{-\mu t}\|\eta_0\|_2 + \int_0^t e^{-\mu(t-s)}e^{-\mu(T-s)}\|\eta(s)\|_2 ds$$

$$+c \int_0^t e^{-\mu(t-s)} \left\{[cM^2(e^{-2\mu(T-s)} + e^{-2\mu s})] + [\|\varphi_0\|_2 + cM^2]e^{-\mu(T-s)}\right\} ds .$$

We apply Gronwall lemma to conclude that

$$\|\eta^T(t)\|_2 \le c\,[\|\eta_0\|_2 + \|\varphi_0\|_2 + c\,M^2](e^{-\mu(T-t)} + e^{-\mu t}) .$$

Now, from the equality $\varphi = \mathscr{E}(T-t)\eta + h^T$, we deduce that a similar estimate holds for φ, namely,

$$\|\varphi^T(t)\|_2 \le c\,[\|\eta_0\|_2 + \|\varphi_0\|_2 + c\,M^2](e^{-\mu(T-t)} + e^{-\mu t}) .$$

We go back again on the first equation, observing that

$$\eta_t - \Delta\eta = \chi$$

where $\chi := -f'(\bar{y})\eta - \varphi\chi_\omega + R_1(\hat{\eta})$ satisfies

$$\|\chi(t)\|_2 \le c\left\{[\|\eta_0\|_2 + \|\varphi_0\|_2 + c\,M^2](e^{-\mu(T-t)} + e^{-\mu t})\right\} \tag{27}$$

Since in dimension $n \le 3$ we have $2 > \frac{n}{2}$, this implies an estimate for η in L^∞, because, as is well-known, the heat semigroup yields

$$\|\eta(t)\|_\infty \le c\left\{\|\chi\|_{L^\infty((t-1,t);L^r(\Omega))} + \|\eta\|_{L^2((t-1,t)\times\Omega)}\right\}$$

for $r > \frac{n}{2}$. Therefore, we conclude that

$$\|\eta(t)\|_\infty \le c\left\{[\|\eta_0\|_2 + [\|\varphi_0\|_2 + c\,M^2](e^{-\mu(T-t)} + e^{-\mu t})\right\} ,$$

for $t \ge 1$ and, if $\eta_0 \in L^\infty(\Omega)$, the estimate extends to $[0,1]$ as well (with a constant now depending on $\|\eta_0\|_\infty$). Similarly we reason for φ; finally, we proved that

$$\|\eta(t)\|_\infty + \|\varphi(t)\|_\infty \le c[\|\eta_0\|_\infty + \|\varphi_0\|_\infty + cM^2](e^{-\mu(T-t)} + e^{-\mu t}) .$$

Choose now some $M \le 1$ such that $cM^2 \le \frac{M}{2}$; then, if $\|\eta_0\|_\infty + \|\varphi_0\|_\infty$ are suitably bounded, we have

$$c[\|\eta_0\|_\infty + \|\varphi_0\|_\infty + cM^2] \le M$$

so that X becomes an invariant convex subset of $L^2(0, T; L^2(\Omega))$. Continuity and compactness of the operator $\mathcal{K}$ are easy to prove, which allows us to conclude the existence of a fixed point (η, φ) which is therefore a solution to (18).   □

3.2 Small Solutions

As mentioned above, in this section we consider the particular case where both the target z and the initial datum y_0 are small in $L^2(\Omega)$. In this case, one can prove that the optimal pair for the steady-state problem is unique and the linearized optimality system exponentially stable. The uniqueness of the optima could be expected for the evolution problem as well. This would yield the uniqueness of the solution of the optimality system and would allow to apply the previous result. But the arguments of the steady-state case only allow proving the smallness of the time-averages of the optimal pairs and this is not sufficient by now to prove the strict convexity of the functional and to conclude.

We consider first the elliptic problem:

$$\min \{J(u) := \frac{1}{2}|y - z|^2_{L^2(\omega_0)} + \frac{1}{2}|u|^2_{L^2(\omega)}\} ,$$ (28)

associated to

$$\begin{cases} -\Delta y + f(y) = u\chi_\omega & \text{in } \Omega \\ y = 0 & \text{on } \partial\Omega . \end{cases}$$ (29)

Obviously

$$I = \min J \le J(0) = \frac{1}{2}|z|^2_{L^2(\omega_0)}.$$

Consequently, any minimizer $(\bar{u}, \bar{y})$ satisfies

$$|\bar{y} - z|^2_{L^2(\omega_0)} + |\bar{u}|^2_{L^2(\omega)} \le |z|^2_{L^2(\omega_0)}.$$

Now, assuming that the target z is small enough, this ensures necessarily the smallness of the optimal control $\bar{u}$ and of the optimal state $\bar{y}$, that consequently live in a ball B in $L^2(\omega) \times L^2(\omega_0)$. The radius of this ball, centered at the origin, can be made small as the norm of z in $L^2(\Omega)$ tends to zero. Let us now explain why,

z being small, the functional J is strictly convex in the relevant ball B. In view of Proposition 2.3 in [5] (see also [6] and [21]) we have

$$J''(u)v_1 v_2 = \int_{\omega_0} \eta_{v_1} \eta_{v_2} dx + \int_{\omega} v_1 v_2 dx - \int_{\Omega} f''(y)q\eta_{v_1}\eta_{v_2}, \tag{30}$$

where q is the adjoint state solution of

$$\begin{cases} -\Delta q + f'(y)q = (y-z)\chi_{\omega_0} & \text{in } \Omega \\ q = 0 & \text{on } \partial\Omega, j = 1, 2. \end{cases} \tag{31}$$

and η_{v_1}, η_{v_2} are the linearized solutions in the direction of v_1 and v_2 respectively, i.e.

$$\begin{cases} -\Delta \eta_{v_j} + f'(y)\eta_{v_j} = v_j \chi_\omega & \text{in } \Omega \\ \eta_{v_j} = 0 & \text{on } \partial\Omega, j = 1, 2. \end{cases} \tag{32}$$

Now, assuming that the target z is small enough in $L^2(\Omega)$, we can deduce that both y and q are small in $L^\infty(\Omega)$. Indeed, since $n \leq 3$, the smallness for the control in $L^2(\omega)$ implies that the right-hand side of (29) is small in some $L^p(\Omega)$ with $p > \frac{n}{2}$. Since the nonlinearity is accretive (i.e. it has a good sign), the elliptic regularity implies the smallness of the state in $H_0^1(\Omega) \cap L^\infty(\Omega)$. In turn, a similar property holds for q from (31), in view of the fact that y is bounded and $f'(y) \geq 0$. Therefore, since

$$\|\eta_v\|_{H_0^1(\Omega)} \leq C\|v\|_{L^2(\omega)}$$

the last term in (30) can be absorbed in the second one thanks to the bound of y and the smallness of q in $L^\infty(\Omega)$, namely

$$J''(u)vv \geq \int_{\omega_0} \eta_v^2 dx + (1 - c\|q\|_\infty) \int_\omega v^2 dx .$$

This guarantees the uniqueness of the minimizer of J, but also the uniqueness of a critical point of J in the ball B. Accordingly, one can guarantee that the unique solution of the stationary optimality system on that ball is the minimizer $\bar{u}$. In addition, as shown above, $J''(\bar{u})$ turns out to be coercive, which implies that the linearized optimality system (21) satisfies the exponential turnpike property in this case and Theorem 1 can be applied producing a turnpike solution of the optimality system. Whether the uniqueness of the minimizer is true in the parabolic case and, consequently, if the turnpike property actually holds for the optima under smallness conditions on the initial datum and target is an interesting open problem.

4 Time Independent Controls by Γ-Convergence

Let us consider again the semilinear heat equation

$$
\begin{cases}
y_t - \Delta y + f(y) = u(x)\chi_\omega & \text{in } \Omega \times (0, T) \\
y = 0 & \text{on } (0, T) \times \partial\Omega \\
y(0) = y_0 \in L^2(\Omega) ,
\end{cases}
\tag{33}
$$

but this time with controls $u = u(x)$ independent of time.
     We focus in the particular case:

$$
\begin{cases}
y_t - \Delta y + |y|^{p-1}y = u(x)\chi_\omega & \text{in } \Omega \times (0, T) \\
y = 0 & \text{on } (0, T) \times \partial\Omega \\
y(0) = y_0 \in L^2(\Omega) ,
\end{cases}
\tag{34}
$$

with $p > 1$. We now consider the optimal control problem:

$$
\min \ J^T(u) = \frac{1}{2} \int_0^T |y(t) - z|^2_{L^2(\omega_0)} dt + \frac{T}{2}|u|^2_{L^2(\omega)} ,
\tag{35}
$$

and the steady state version

$$
\begin{cases}
-\Delta y + |y|^{p-1}y = u\chi_\omega & \text{in } \Omega \\
y = 0 & \text{on } \partial\Omega ,
\end{cases}
\tag{36}
$$

together with the corresponding functional

$$
\min \ J(u) = \frac{1}{2}\left[|u|^2_{L^2(\omega)} + |y - z|^2_{L^2(\omega_0)}\right] .
\tag{37}
$$

     Employing Γ-convergence arguments and taking advantage of the fact that the controls under consideration are independent of t the following can be proved.

Theorem 2 *Let u^T be a family of optimal controls for (34) and (35), with $T \to \infty$. Then, this family is relatively compact in $L^2(\omega)$ and any accumulation point $\bar{u}$ as $T \to \infty$ is an optimal control for the steady state problem (36) and (37).*

Remark 2 Note that the uniqueness of the optimal control is not guaranteed nor for the time-dependent problem nor for the steady state one. This is due to the lack of convexity of the functionals under minimization which is derived from the nonlinear character of the state equations. Thus, the statement above necessarily refers to the accumulation points of the family u^T as T tends to infinity and its inclusion within the set of steady state controls.

Proof We proceed in several steps.

Step 1. Let I^T and I be the values of the minimizers for the time-dependent problem in $[0, T]$ and the steady state one. We claim that

$$\frac{I^T}{T} \leq I + O(T^{-1}).$$

To prove this first estimate on the comparison between I and I^T we take a minimizer $\bar{u}$ for I and plug it into the functional J^T. We have

$$\frac{J^T(\bar{u})}{T} - J(\bar{u}) = \frac{\int_0^T |y^T(t) - z|^2_{L^2(\omega_0)} dt}{2T} - \frac{1}{2}|\bar{y} - z|^2_{L^2(\omega_0)}. \tag{38}$$

Here $\bar{y}$ stands for the steady state solution associated to the optimal control $\bar{u}$ and y^T is the corresponding solution of the evolution problem in the interval $[0, T]$. Obviously, because of the monotone character of the nonlinearity, standard energy estimates lead to the following exponential convergence property:

$$||y^T(t) - \bar{y}||_{L^2(\Omega)} \leq \exp(-\lambda_1 t)||y_0 - \bar{y}||_{L^2(\Omega)}, \tag{39}$$

λ_1 being the first eigenvalue of the Dirichlet Laplacian. Obviously, in view of this, the right hand side of (38) can be estimated as $O(T^{-1})$.

Step 2. Similarly, we may prove that

$$I \leq \frac{I^T}{T} + O(T^{-1}).$$

To prove it we proceed all the way around. We plug the minimizer u^T of J^T into the functional J. We have

$$I \leq J(u^T) = \frac{J^T(u^T)}{T} - \frac{\int_0^T ||y^T(t) - z||^2_{L^2(\omega_0)}}{T} + ||\bar{y}^T - z||^2_{L^2(\omega_0)}. \tag{40}$$

This time, y^T stands for the solution of the evolution problem corresponding to the optimal control u^T while $\bar{y}^T$ is the corresponding steady state solution.

From previous estimates we know that

$$||y^T(t) - \bar{y}^T||_{L^2(\Omega)} \leq \exp(-\lambda_1 t)||y_0 - \bar{y}^T||_{L^2(\Omega)}, \tag{41}$$

so that, in order to guarantee uniform (with respect to T) decay rates, we need $\bar{y}^T$ to be uniformly bounded in $L^2(\Omega)$ and this requires uniform bounds on u^T in $L^2(\omega)$. But the uniform bound on u^T is easy to achieve.

Indeed, the solution of the evolution problem with $u = 0$ decays exponentially to zero. Thus, $\frac{1}{T}J^T(0)$ is uniformly bounded as T tends to infinity. Consequently

I^T/T is bounded above and this yields to the uniform bound of u^T in $L^2(\Omega)$. This automatically leads to uniform estimates for $\bar{y}^T$ in $H_0^1(\Omega) \cap L^{p+1}(\Omega)$. Therefore, (41) implies

$$||y^T(t) - \bar{y}^T||_{L^2(\Omega)} \leq c \exp(-\lambda_1 t) ,$$

and thanks to this estimate we have

$$||\bar{y}^T - z||^2_{L^2(\omega_0)} - \tfrac{1}{T} \int_0^T ||y^T(t) - z||^2_{L^2(\omega_0)}$$

$$\leq \tfrac{1}{T} \int_0^T [||\bar{y}^T - z||^2_{L^2(\omega_0)} - ||y^T(t) - z||^2_{L^2(\omega_0)}]dt = O(T^{-1}) .$$

Therefore from (40) we conclude that $I \leq \frac{I^T}{T} + O(T^{-1})$.

Step 3. Let us now show that the accumulation points of a sequence of minimizers u^T of J^T as T tends to infinity, are necessarily minimizers of J. The sequence u^T being bounded in $L^2(\omega)$, by extracting subsequences, we can get a weak limit in $L^2(\omega)$ that we denote as u^*. We claim that u^* is a minimizer for J. In other words, that $J(u^*) = I$. To prove it, we compute $J(u^*)$. We claim that, using lower semicontinuity properties, it follows that

$$J(u^*) \leq \liminf_{T \to \infty} \frac{J^T(u^T)}{T} . \tag{42}$$

Furthermore, according to the results in Steps 1 and 2, it follows that

$$\liminf_{T \to \infty} \frac{J^T(u^T)}{T} = \liminf_{T \to \infty} \frac{I^T}{T} = \lim_{T \to \infty} \frac{I^T}{T} = I .$$

This implies that $J(u^*) \leq I$ and, consequently, u^* is a minimizer for the steady state problem.

Let us now prove the claim (42). Obviously, by the weak convergence of the controls we have

$$||u^*||_{L^2(\omega)} \leq \liminf_{T \to \infty} ||u^T||_{L^2(\omega)} . \tag{43}$$

We need also to compare the solution $\bar{y}^*$ of the steady state problem associated to u^* and the solution y^T associated to u^T, the chosen optimal pairs for the time evolution problem in the time intervals $[0, T]$. We claim that $[\int_0^T y^T dt]/T$ converges to $\bar{y}^*$ in $L^2(\Omega)$.

Indeed, from previous estimates we know the uniform boundedness of $\bar{y}^T$ in $H_0^1(\Omega) \cap L^{p+1}(\Omega)$, where $\bar{y}^T$ is the steady state solution associated to u^T. Passing to the limit in the steady state problem it can be shown easily that $\bar{y}^T$ weakly converges in $H_0^1(\Omega) \cap L^{p+1}(\Omega)$ to $\bar{y}^*$. Since we have

$$||\frac{\int_0^T y^T(t)dt}{T} - \bar{y}^*||_{L^2(\Omega)} \le ||\frac{\int_0^T y^T(t)dt}{T} - \bar{y}^T||_{L^2(\Omega)} + ||\bar{y}^T - \bar{y}^*||_{L^2(\Omega)} ,$$

taking into account the fact that

$$||\bar{y}^T - \bar{y}^*||_{L^2(\Omega)} \to 0 ,$$

as T tends to ∞, and that, due to (41),

$$||\frac{\int_0^T y^T(t)dt}{T} - \bar{y}^T||_{L^2(\Omega)} = O(T^{-1}) ,$$

we conclude that

$$\frac{\int_0^T y^T(t)dt}{T} \to \bar{y}^* \quad \text{in} \quad L^2(\Omega) \text{ as } T \to \infty .$$

Finally, we have

$$\frac{J^T(u^T)}{T} \ge \frac{1}{2}\left\{ ||\frac{1}{T}\int_0^T y^T dt - z||_{L^2(\Omega)}^2 + ||u^T||_{L^2(\Omega)}^2\right\}$$

and with the convergence established above we complete the proof of the claim (42).

So far we have proved the weak convergence in L^2 of the controls u^T towards a limit control u^*. But the arguments of Step 1 and 2, showing that $\frac{1}{T}I^T \to I$, imply that the norms also converge. This leads to strong convergence as stated in the Theorem.   □

Remark 3 Several remarks are in order:

- The result above, employing Γ-convergence, does not use the controllability properties of the system but only its exponential stability as time tends to infinity.
- The proof above does not yield any convergence rates.

5 Further Comments and Open Problems

1. The Γ-convergence proof above uses in an essential way the fact that the controls under consideration are independent of time.

   It would be very interesting to analyze whether these techniques can be applied for time-evolving controls.

   In this argument we use standard stability properties of the semilinear heat equation with nonlinearities satisfying the good sign condition. Of course this proof can be generalized for a larger class of semilinear problems enjoying properties similar to (39). These techniques can also be employed, for instance, for damped semilinear wave equations.

2. It would be interesting to investigate similar questions for more general nonlinearities, leading possibly to more complex dynamics, such as:

$$y_t - \Delta y + y^3 - Ly = u\chi_\omega \quad \text{in } \Omega \times (0, \infty) \tag{44}$$

for $L > 0$. Note that, indeed, when $L > \lambda_1$, the first eigenvalue of the Dirichlet laplacian, the existence of a steady state solution is guaranteed but not its uniqueness. Also, the trajectories of the parabolic problem are bounded and the Lyapunov function

$$\frac{1}{2}\int_\Omega [|\nabla y|^2 - Ly^2]dx + \frac{1}{4}\int_\Omega y^4 dx - \int_\omega uy dx$$

allows proving that all the elements of the ω-limit set are steady state solutions.

In the present case, according to the results in [13] and the analyticity of the nonlinearity one can prove that every trajectory converges as $t \to \infty$ to a steady state solution. Whether this kind of results and the techniques in [13] can be further developed to obtain results of turnpike nature is an interesting open problem.

3. The results we have obtained in Sect. 3 apply to a particular class of solutions of the optimality system. But, as mentioned above, even in the case of small initial data and target, further work is needed to show its applicability to the optimal pairs.

In any case, the obtained results are of local nature. The obtention of global results would require a more complete understanding of the controlled dynamics. This has been done successfully in a number of examples (see [3] and [4]). But a systematic approach to these problems is still to be developed.

Note in particular that the optimal control and states are not unique for the steady state problem under consideration. Actually, the multiplicity of its solutions, its stability properties and the impact this might have on the problem under consideration could be worth investigating.

4. It would be natural to consider the analogues of the optimal control problems above in the context of optimal shape design. Often in applications, in particular in aeronautics (see [12]), optimal shapes are computed on the basis of the steady state modeling but they are then employed in time evolving ones. In the particular context of the semilinear heat equation of the previous section, the following problem makes sense. Does the optimal shape in the time interval $[0, T]$ converge, as T tends to infinity, to the optimal steady state shape for the elliptic equation? Of course this problem can be formulated in a variety of contexts, depending mainly on the admissible class of shapes considered.

   There is very little in the subject except for the paper [1] where the issue is addressed in the context of the two-phase optimal design of the coefficients. In this setting it is indeed proved that, as time tends to infinity, optimal designs of the parabolic dynamics converge to those of the elliptic steady-state problem.

5. Note that the problems of optimal shape design and its possible stabilization in long time horizons could be much more complex if one would consider, for the evolution problem, shapes allowed to evolve in time as well.

6. At this stage it is important to observe that, according to earlier results in [16] and [17] for the conservative wave equations, in the absence of damping, optimal shape design problems for the collocation of actuators and sensors in long time intervals lead to spectral problems, and not really to steady state ones. In other words, for conservative dynamics one does not expect the simplification of optimal design problems to occur towards the steady state optimal design problem, but rather towards a spectral version of it. The results in [19] for the heat equation show that the optimal designs are determined by a finite number of eigenfunctions that diminishes as the time horizon increases. Thus even in this parabolic context, the relevant optimal design problem is not a steady state one but of spectral nature even if it involves only a finite number of eigenfunctions. Thus, the overall role of the turnpike property for shape design problems, in particular in the context of optimal placement of sensors and actuators, is to be clarified.

7. At this point it is worth to underline that, while in [1] the authors work with given initial data, in the problem considered in [19], the optimal placement of sensors and actuators is determined within the whole class of solutions. The questions are then of different nature and, accordingly, the expected results are not necessarily the same. This issue is important for a correct formulation of the optimal design problem to be addressed and the comparison of the corresponding results.

   Note also that the reduction in [19] requires of a randomization procedure so that optimal shapes are defined to be optimal in some probabilistic sense. But, roughly, it can be said that the results above apply in the context where the optimal shape of the actuator or sensor is computed so to be optimal within the whole class of solutions of the PDE under consideration. Of course, all these problems are expected to be easier to handle when one considers given fixed initial data as pointed out in [18], which corresponds, somehow, to the situation considered in [1].

8. The same can be said about inverse problems. The problem of the connection between the inversion process in long time intervals and in the steady state regime makes fully sense both in the context of linear and nonlinear problems and in a variety of inverse problems. Very little is known in this subject (see [11], section 9.4).

Acknowledgements Enrique Zuazua was partially supported by the Advanced Grant NUMERIWAVES/FP7-246775 of the European Research Council Executive Agency, the FA9550-14-1-0214 of the EOARD-AFOSR, FA9550-15-1-0027 of AFOSR, the MTM2011-29306 and MTM2014-52347 Grants of the MINECO, and a Humboldt Award at the University of Erlangen-Nürnberg. This work was done while the second author was visiting the Laboratoire Jacques Louis Lions with the support of the Paris City Hall "Research in Paris" program.

References

1. Allaire, G., Münch, A., Periago, F.: Long time behavior of a two-phase optimal design for the heat equation. SIAM J. Control. Optim. **48**, 5333–5356 (2010)
2. Anderson, B.D.O., Kokotovic, P.V.: Optimal control problems over large time intervals. Autom. J. IFAC **23**, 355–363 (1987)
3. Cardaliaguet, P., Lasry, J-M., Lions, P.-L., Porretta, A.: Long time average of mean field games. Netw. Heterog. Media **7**, 279–301 (2012)
4. Cardaliaguet, P., Lasry, J-M., Lions, P.-L., Porretta, A.: Long time average of mean field games in case of nonlocal coupling. SIAM J. Control. Optim. **51**, 3558–3591 (2013)
5. Casas, E., Mateos, M.: Optimal Control for Partial Differential Equations. Proccedings of Escuela Hispano Francesa 2016. Oviedo, Spain (to appear)
6. Casas, E., Tröltzsch, F.: Second order analysis for optimal control problems: improving results expected from abstract theory. SIAM J. Optim. **22**, 261–279 (2012)
7. Chichilnisky, G.: What is Sustainable Development? Man-Made Climate Change, pp. 42–82. Physica-Verlag HD, Heidelberg/New York (1999)
8. Choulli, M.: Une introduction aux problèmes inverses elliptiques et paraboliques. Mathematiques & Applications, vol. 65. Springer, Berlin (2009)
9. Damm, T., Grüne, L., Stielerz, M., Worthmann, K.: An exponential turnpike theorem for dissipative discrete time optimal control problems. SIAM J. Control. Optim. **52**, 1935–1957 (2014)
10. Grüne, L.: Economic receding horizon control without terminal. Autom. J. IFAC **49**, 725–734 (2013)
11. Isakov, V.: Inverse Problems for Partial Differential Equations, 2nd edn. Applied Mathematical Sciences, vol. 127. Springer, New York (2006)
12. Jameson, A.: Optimization methods in computational fluid dynamics (with Ou, K.). In: Blockley, R., Shyy, W. (eds.) Encyclopedia of Aerospace Engineering. John Wiley & Sons, Hoboken (2010)
13. Jendoubi, M.A.: A simple unified approach to some convergence theorems of L. Simon. J. Funct. Anal. **153**, 187–202 (1998)
14. Nodet, M., Bonan, B., Ozenda O., Ritz, C.: Data Assimilation in Glaciology. Advanced Data Assimilation for Geosciences. Les Houches, France (2012)
15. Porretta A., Zuazua, E.: Long time versus steady state optimal control. SIAM J. Control. Optim. **51**, 4242–4273 (2013)
16. Privat, Y., Trélat, E., Zuazua, E.: Optimal location of controllers for the one-dimensional wave equation. Ann. Inst. H. Poincaré Anal. Non Linéaire **30**, 1097–1126 (2013)
17. Privat, Y., Trélat, E., Zuazua, E.: Optimal observation of the one-dimensional wave equation. J. Fourier Anal. Appl. **19**, 514–544 (2013)
18. Privat, Y., Trélat, E., Zuazua, E.: Complexity and regularity of maximal energy domains for the wave equation with fixed initial data. Discret. Cont. Dyn. Syst. **35**, 6133–6153 (2015)
19. Privat, Y., Trélat, E., Zuazua, E.: Optimal shape and location of sensors and controllers for parabolic equations with random initial data. Arch. Ration. Mech. Anal. **216**, 921–981 (2015)
20. Trélat, E., Zuazua, E.: The turnpike property in finite-dimensional nonlinear optimal control. J. Differ. Equ. **258**, 81–114 (2015)

21. Tröltzsch, F.: Optimal Control of Partial Differential Equations. Theory, Methods and Applications. Graduate Studies in Mathematics, vol. 112. American Mathematical Society, Providence (2010)
22. Zaslavski, A.J.: Turnpike properties in the calculus of variations and optimal control. Nonconvex Optimization and its Applications, vol. 80. Springer, New York (2006)
23. Zuazua, E.: Propagation, observation, and control of waves approximated by finite difference methods. SIAM Rev. **47**(2), 197–243 (2005)
24. Zuazua, E.: Controllability and observability of partial differential equations: some results and open problems. In: Dafermos, C.M., Feireisl E. (eds.) Handbook of Differential Equations: Evolutionary Equations, vol. 3, pp. 527–621. Elsevier Science, Amsterdam/Boston (2006)

Part III
Paleoclimate

Effects of Additive Noise on the Stability of Glacial Cycles

Takahito Mitsui and Michel Crucifix

Abstract It is well acknowledged that the sequence of glacial-interglacial cycles is paced by the astronomical forcing. However, how much is the sequence robust against natural fluctuations associated, for example, with the chaotic motions of atmosphere and oceans? In this article, the stability of the glacial-interglacial cycles is investigated on the basis of simple conceptual models. Specifically, we study the influence of additive white Gaussian noise on the sequence of the glacial cycles generated by stochastic versions of several low-order dynamical system models proposed in the literature. In the original deterministic case, the models exhibit different types of attractors: a quasiperiodic attractor, a piecewise continuous attractor, strange nonchaotic attractors, and a chaotic attractor. We show that the combination of the quasiperiodic astronomical forcing and additive fluctuations induces a form of temporarily quantised instability. More precisely, climate trajectories corresponding to different noise realizations generally cluster around a small number of stable or transiently stable trajectories present in the deterministic system. Furthermore, these stochastic trajectories may show sensitive dependence on very small amounts of perturbations at key times. Consistently with the complexity of each attractor, the number of trajectories leaking from the clusters may range from almost zero (the model with a quasiperiodic attractor) to a significant fraction of the total (the model with a chaotic attractor), the models with strange nonchaotic attractors being intermediate. Finally, we discuss the implications of this investigation for research programmes based on numerical simulators.

T. Mitsui
Earth and Life Institute, George Lemaître Centre for Earth and Climate Research,
Université catholique de Louvain, BE-1348 Louvain-la-Neuve, Belgium
e-mail: takahito321@gmail.com

M. Crucifix (✉)
Earth and Life Institute, George Lemaître Centre for Earth and Climate Research,
Université catholique de Louvain, BE-1348 Louvain-la-Neuve, Belgium

Belgian National Fund of Scientific Research, Rue d'Egmont, 5 BE-1000 Brussels, Belgium
e-mail: michel.crucifix@uclouvain.be

© Springer International Publishing Switzerland 2016
F. Ancona et al. (eds.), *Mathematical Paradigms of Climate Science*, Springer
INdAM Series 15, DOI 10.1007/978-3-319-39092-5_6

Keywords Ice ages • Milankovitch • Strange nonchaotic attractor • Dynamical systems • Van der Pol oscillator

1 Introduction

Analyses of marine sediments and ice core records show, among others, that glacial and interglacial periods alternated over the last three million years [44, 64]. These are major climate changes. The last glacial maximum that occurred about 21,000 years ago was characterised by extensive ice sheets over large fractions of North America, the British Isles and Fennoscandia. Sea-level was about 120 m below the present-day, and the CO_2 concentration was about 90 ppm lower than its typical pre-industrial value of 280 ppmv [41, 55].

The so-called "LR04" time series [44] is a compilation of records of Oxygen isotopic ratio $\delta^{18}O$ recorded in deep-sea organisms. This is representative of the succession of glacial-interglacial cycles (Fig. 1). It may be seen that the temporal signature of glacial-interglacial cycles has evolved through time: their amplitude increased gradually, and about 1 million years ago, their period settled to about 100 ka.[1] The four latest cycles are particularly distinctive, with a gradual glaciation phase extending over about 80 ka, and a deglaciation over 10–20 ka [13, 55].

Glacial cycles emerge from a complex interplay of various physical, biogeo-chemical and geological processes, and it is hoped that their detailed analysis will yield information on the stability of the different components of the climate system. It has also become clear that glacial cycles are submitted to an external control. In particular, the timing of deglaciations is statistically related to quasiperiodic changes in Earth's orbit and obliquity [34, 43, 58]. One of the key mechanisms of this control is that changes in Earth's orbit and obliquity influence the seasonal and spatial distributions of the incoming solar radiation (insolation) at the top of the atmosphere. Specifically, the insolation of at a given time of the year at a given latitude is approximately a linear function of $e \sin \varpi$, $e \cos \varpi$, and ε, where e is the Earth orbit eccentricity, ϖ the longitude of the perihelion, and ε the Earth's obliquity. The quantity $e \sin \varpi$ is sometimes called the climatic precession parameter [9]. In turn, summer insolation at high latitude controls the mass balance of snow over the years, and affects thus the growth of ice sheets [1, 6, 45, 70]. This is, however, probably not the only mechanism of astronomical control on ice ages [61].

A long-standing puzzling fact is that the Fourier spectrum of insolation changes mainly contains power around 20 and 40 ka [11], while the spectrum of the slow fluctuations of climate shows a concentration of power around 100 ka (Fig. 1, see also [12, 32, 59, 71]). The periods of 19–23 ka arise from precession (the rotation period of ϖ) while 40 ka is the dominant period of obliquity. In fact, periodicities around 100 ka do appear in astronomical forcing, but somewhat indirectly [10].

---

[1]In the following, 1 ka = 1,000 years and 1 Ma = 1,000 ka.

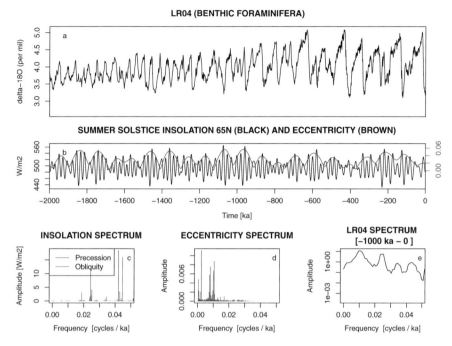

Fig. 1 (**a**) Reconstructed climate variations over the last 2 million years (1 ka stands for 1,000 years) inferred from deep-sea organisms, specifically benthic foraminifera [44], along with (**b**) the variations in incoming solar radiation at the summer solstice at 65° N (*black*), a classical measure of astronomical forcing computed here following the BER90 algorithm [7]. The spectrum of insolation is (**c**), with components arising from climatic precession and obliquity computed following [8, 11]. Eccentricity (figure (**b**), *brown*) is the modulating envelope of precession, and its spectrum is given in (**d**) [8, 11]. (**e**) multi-taper estimate of the LR04 spectrum (last 1 million years only) estimated using multi-taper method [66] obtained using the SSA-MTM toolkit [69] with default parameters

In particular, eccentricity, which modulates the amplitude of climatic precession, is characterised by a spectrum with periods around 100 and 413 ka [5, 7]. The correspondence between the 100 ka period of eccentricity and the duration of ice ages was noted early on [32, 35], and statistical analysis of the timing of ice ages indicates that this correspondence is probably not fortuitous [35, 43, 58].

These observations lead us to an interesting problem: is it possible to predict the effects of the astronomical forcing on the ice ages without full knowledge of the detailed physical mechanisms? Specifically, is it possible to determine whether the sequence of ice ages is tightly controlled by the astronomical forcing or whether, to the contrary, this sequence is highly sensitive to small fluctuations?

The strategy proposed here relies on the analysis of low-order dynamical systems. Over the years, numerous models have been proposed to explain, on the one hand, the relationship between ice ages and astronomical forcing, and, on the other hand, their specific saw-tooth temporal structure. A full review is beyond the scope of the present study, and we quote here some potential dynamical mechanisms, which may be relevant for modelling of ice ages:

(a) Considerations on the geometry of ice sheets [70] suggest that a positive insolation anomaly may be proportionally more effective than a negative one. In terms of system dynamics, one may say that the forcing is transformed nonlinearly by the climate system (see also Fig. 8 of [61] for other nonlinear effects). Recall that precession exerts a significant control on the seasonal distribution of insolation, and that the modulating envelope of the precession signal is eccentricity. A nonlinear transformation of the astronomical forcing is thus a simple mechanism by which the spectrum of eccentricity can make its way towards the spectrum of climate variations. The spectrum of eccentricity includes the sought-after 100-ka period, but it is also dominated by a 413-ka period, and the latter is not observed in the benthic record. This was sometimes referred to as the 400-ka enigma [26, 59].

(b) Results from numerical modelling suggest that the ice-sheet-atmosphere system may present several stable states for a range of insolation forcings [1, 14, 15, 70]. Transitions between these states may be triggered deterministically (by the forcing), stochastically, or by a combination of both. In early studies [4, 48], it was suggested that 100-ka ice ages may emerge through a mechanism of stochastic resonance, in which noisy fluctuations may amplify the small eccentricity signal. This proposal is relevant to the present context because it is the first one conferring an explicit role to fluctuations, but the stochastic resonance theory of ice ages is incomplete because it does not consider explicitly the direct effects of precession and obliquity, nor does it explain the saw-tooth shape of glacial cycles.

(c) The 100 ka cycles may arise as self-sustained (or excitable) oscillations, which emerge from nonlinear interactions between different Earth system components. These components can be ice sheets (and underlying lithosphere), deep-ocean and carbon cycle dynamics. Following this approach, the effect of astronomical forcing on climate may be understood in terms of the general concept of synchronisation [3, 19, 68], of which the forced van der Pol oscillator may constitute a paradigm. Saltzman et al. [62, 63] published a number of low-order dynamical systems consistent with this interpretation, but see also [3, 27, 52] for alternatives.

(d) If a process of nonlinear resonance occurs, 100-ka glacial cycles may also be obtained even if the corresponding autonomous system does not have any internal period near 100 ka. A classical example of nonlinear resonance is the Duffing oscillator [37], which was recently suggested as a possible basis for the investigation of ice age dynamics [18]. While, in principle, nonlinear resonance with additive forcing may be sufficient to generate combination of tones (see also [42]), the expression of a dominant 100-ka cycle in response to the astronomical forcing is best obtained with multiplicative forcing [18, 33, 42].

(e) Ice age cycles may also be obtained in a more ad-hoc way: for example by resorting to a discrete variable that changes states following threshold rules involving the astronomical forcing [51], or by postulating an adequate bifurcation structure in the climate-forcing space [21].

Naturally, different models exhibit different dynamical properties. On this subject, it was observed that models explaining ice ages as the result of self-sustained oscillations subjected to the astronomical forcing generally display the properties of strange nonchaotic attractors [17, 46, 47].

Strange nonchaotic attractors (SNAs) may appear when nonlinear dynamical systems are forced by quasiperiodic signals [28, 38], such as the astronomical forcing. Unlike chaotic systems, the trajectory of such systems is typically robust against small fluctuations in the initial conditions. In particular, their largest Lyapunov exponent λ is nonpositive. However, the attractor itself, or its stroboscopic section at one of the periodic components of the forcing, is a geometrically strange set. Comprehensive reviews on SNAs are available in [23, 56]. The strange geometry of SNAs is related to the existence of repellers of measure zero embedded in the attractor [65]. Thus, there will be times at which the orbits on SNAs are arbitrarily close to the orbits on the repellers. As a result, it is shown that the trajectories generated by models with SNAs may have sensitive dependence on parameters [50]. They may also show sensitive dependence on dynamical noise [40].

The concept of SNA was introduced in ice age theory on the basis of low-order deterministic models [17, 46, 47], but such models are naturally gross simplifications of the complex climate system. In this context, one step forward is to enrich the dynamics with stochastic parameterisations, in order to represent the effects of fast climate and meteorological processes [31, 54, 63]. As already mentioned, stochastic parameterisations were introduced in the palaeoclimate context as an element of the stochastic resonance theory [4, 48]. Stochastic processes were also considered in simple ice age models to illustrate the process of synchronisation [15, 49, 68], to induce stochastic jumps between different stable equilibria [21], or to induce coherence resonant oscillations [53].

Our specific objective is here to study the effect of dynamical noise (or system noise) on the robustness of palaeoclimate trajectories generated by simple stochastic models forced by the astronomical forcing, and explain differences among the models by reference to the properties of the attractors displayed by the deterministic counterparts of these models. To this end, we consider four simple models known to exhibit different kinds of attractors, and consider the effects of additive white Gaussian noise on the simulated sequence of ice ages.

Modelling fast meteorological and climatic processes by additive Gaussian noise may be oversimplified though it is frequently used in studies of ice ages [4, 21, 48, 63, 68]. In the studies of millennial-scale climate changes (so-called Dansgaard Oeschger events), Ditlevsen (1999) employs the α-stable noise, which is characterized by a fat-tailed density distribution [20], and mathematical methods to identify the α-stable noise in time series have been developed (see Gairing et al. in this volume). Colored multiplicative noises were also introduced in box models of thermohaline circulation [67]. Such non-Gaussian or colored multiplicative noises can be relevant to ice age dynamics. However, here we focus on the effects of additive white Gaussian noises as a first step to examine the stability of ice age models.

2   Methods

The astronomical forcing $F(t)$ is represented here as a linear combination of forcing functions associated with obliquity and climatic precession. For consistency with previous works we use the summer-solstice standardised (zero-mean) insolation approximated as a sum of 35 periodic functions of time t, as in [19] and [46] (see also [15]),

$$F(t) = \frac{1}{A} \sum_{i=1}^{35} [s_i \sin \omega_i t + c_i \cos \omega_i t], \tag{1}$$

where the scale factor A is set to $11.77\,\text{W/m}^2$ for the CSW model mentioned below and $23.58\,\text{W/m}^2$ for the other models.

The current study is focused around four previously published conceptual models whose parameter values are listed in Appendix:

- The model introduced by Imbrie and Imbrie [36] (I80) is one dimensional ordinary differential equation in which the climate-state x (a measure of the global ice volume loss) responds to the astronomical forcing $F(t)$ as follows:

$$\tau \frac{dx}{dt} = \begin{cases} (1+b)(F(t)-x) & \text{if } F(t) \geq x \\ (1-b)(F(t)-x) & \text{if } F(t) < x. \end{cases} \tag{2}$$

  The additional condition $x \geq 0$ expressed in [36] is omitted in this study for simplicity.
- The P98 model [51] is a hybrid dynamical system defined as follows:

$$\frac{dx}{dt} = \frac{x_R - x}{\tau_R} - \frac{\tilde{F}(t)}{\tau_F}, \tag{3}$$

  where x is the global ice volume, and the relaxation time τ_R and the relaxed state x_R vary discretely between $R = i$, $R = g$, and $R = G$ according to the following transition rules: the transition $i \to g$ is triggered when $F(t)$ falls below a threshold i_0; the transition $g \to G$ is triggered with x exceeds $x_{\max}$, and $G \to i$ when $F(t)$ exceeds a threshold i_1. Note also that the forcing function used in Eq. (3) is a truncated version of actual insolation (a nonlinear effect), computed as follows:

$$\tilde{F}(t) = \frac{1}{2} \left(F(t) + \sqrt{4a^2 + F^2(t)} \right). \tag{4}$$

- The SM90 model [63] is a representation of nonlinear interactions between three components of the Earth system: continental ice volume (x), CO_2 concentration

(y) and deep-ocean temperature (z):

$$\tau \frac{dx}{dt} = -x - y - v\,z - uF(t),$$

$$\tau \frac{dy}{dt} = -p\,z + r\,y + s\,z^2 - w\,yz - z^2 y,$$

$$\tau \frac{dz}{dt} = -q(x + z).$$

The forcing is additive, and the nonlinearity introduced in the second component of the equation induces limit cycle dynamics when $u = 0$.

- The HA02 model: This is the same as SM90, but Hargreaves and Annan (2002) [30] estimated the parameter values of SM90 using a data assimilation technique.
- The CSW model [15, 16, 19] is in fact a forced van der Pol oscillator, used as a simple example of slow-fast oscillator with parameters calibrated such as to reproduce the ice ages record:

$$\tau \frac{dx}{dt} = -\left(\gamma F\left(t\right) + \beta + y\right),$$

$$\tau \frac{dy}{dt} = \alpha(y - y^3/3 + x),$$

where x is the global ice volume, and y is a conceptual variable introduced to obtain a self-sustained oscillation in the absence of forcing ($\gamma = 0$).

Referring to the model categories outlined in Sect. 1, I80 belongs to category (a), SM90, HA04, and CSW to category (c), and P98 to category (e). P98 also has the particularity of being a hybrid dynamical system involving discontinuous thresholds and discrete variables, unlike the other models studied here.

For each model, denote $\mathbf{x}$ the vector of all the climate state variables. The system equations are

$$\frac{d\mathbf{x}}{dt} = \mathbf{f}(\mathbf{x}, F(t)). \tag{5}$$

If we introduce phase variables $\theta_i(t) = \omega_i t \pmod{2\pi}$ ($i = 1, 2, \ldots, 35$), Eq. (5) can be written in a skew-product from:

$$\frac{d\mathbf{x}}{dt} = \tilde{\mathbf{f}}(\mathbf{x}, \theta), \tag{6}$$

$$\frac{d\theta}{dt} = \omega, \tag{7}$$

where $\theta = (\theta_1, \theta_2, \ldots, \theta_{35})$ and $\omega = (\omega_1, \omega_2, \ldots, \omega_{35})$. We consider the attractor of each model in the extended phase space ($\theta, \mathbf{x}$). As time t elapses enough from the initial time t_0, trajectories approach the attractor (cf. [28] for a definition of attractor in this particular context). To see a qualitative difference between the categories, we show the attractors of each model for a simplified forcing $F_s(t) = \frac{1}{A_{\{1,3,4\}}} \sum_{i=1,3,4}(s_i \sin \omega_i t + c_i \cos \omega_i t)$, where indices $i = 1, 3, 4$ and parameter $A_{\{1,3,4\}}$ are consistent with [46]. The attractors of each model for the simplified forcing $F_s(t)$ are shown in Fig. 2. To visualise the high-dimensional attractors, the state points are plotted in a three-dimensional space of $(\theta_3/2\pi, \theta_4/2\pi, x)$ at a regular time interval of $12\pi/\omega_1$ (the so-called stroboscopic section). For each model, these plots show the relationship between the phases of the astronomical forcing and the variable representing ice volume.

The different geometries associated with these models can be readily identified (Fig. 2) (see also [46]). The I80 model has a smooth attractor. More specifically, the

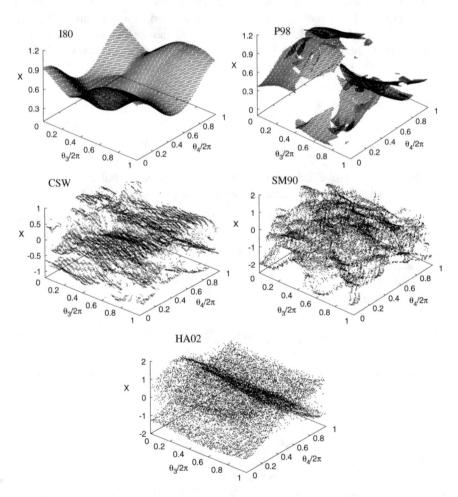

Fig. 2 Attractors of each model for a simplified astronomical forcing $F_s(t)$. To visualise the high-dimensional attractors, the state points are plotted in a three-dimensional space of $(\theta_3/2\pi, \theta_4/2\pi, x)$ at a regular time interval of $12\pi/\omega_1$. Transients are removed

stroboscopic section is a smooth surface, and the time evolution of a trajectory is quasiperiodic. The stroboscopic section of P98 appears as piecewise smooth, which is not surprising considering the fact that the equations are linear, expect for the state transitions at threshold values of state or insolation. Consistently with earlier analysis [46], the CSW and SM90 exhibit SNAs. The stroboscopic sections appear discontinuous almost everywhere. It can qualitatively be discerned that the sections of CSW and SW90 are more organised than the section of HA02, which is known to be chaotic. Recall again that SM90 and HA02 are the same equations, but with different parameters, and the two regimes are separated by a transition from SNA (negative largest Lyapunov exponent λ) to chaos (positive λ).

Stochastic versions are now defined for each model. As ice volume $x(t)$ is the only state variable common among all the models, dynamical noise is added only to the equation of $x(t)$ to allow us to compare models. The equation for the ice volume $x(t)$ is thus schematically written as:

$$dx = f(\mathbf{x}, t)dt + D\,dW(t),$$

where $W(t)$ is the Wiener process, and D is the noise intensity, and $f(\mathbf{x}, t)$ denotes the derivative $\frac{dx}{dt}$ entering the corresponding deterministic model. To account for the fact that the typical size of ice volume variations is different among models, we introduce the scaled noise intensity $\sigma = D/L$, where L is the standard deviation of ice volume $x(t)$ for each model calculated during $[-700\,\text{ka}, 10\,\text{Ma}]$ in the absence of noise (see Appendix for the value of L in each model). The initial time is set at $t_0 = -20\,\text{Ma}$ for SM90 and HA02, and $t_0 = -2\,\text{Ma}$ for the other models to discard initial transients [46]. Each model is integrated from $t = t_0$ to $t_s = -700\,\text{ka}$ without dynamical noise and then integrated with dynamical noise $D\,dW(t)$ from $t = t_s$ to $t = t_e$. We simulate $N(= 200)$ trajectories corresponding to different realizations of the Wiener process $W(t)$. All models are integrated using the stochastic Heun method with a time step of 0.001 ka [29].

The ensemble of ice volumes $\{x_i(t) : i = 1, \ldots, N\}$ disperses due to different noise realizations. Twenty sample trajectories generated by CSW model for $\sigma = 0.002$ are shown in Fig. 3. As earlier noted [17, 19, 46], the astronomical forcing induces a form of synchronisation, such that the different noisy trajectories tend to remain clustered. There are however times at which clusters break apart, yielding a temporarily more disorganised picture. This is the behaviour that we wish to characterise more systematically.

To this end, the dispersions of trajectories in the models are analyzed by using the following three quantities:

- *Size of dispersion of ice volume $x_i(t)$ at a time instant t, $S(t)$, compared to the typical size of ice volume variation L is given by*

$$S(t) = \left(\frac{1}{N}\sum_{i=1}^{N}[x_i(t) - \langle x_i(t)\rangle]^2\right)^{1/2} /L, \qquad (8)$$

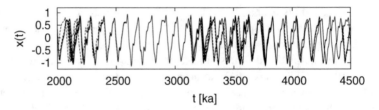

Fig. 3 Effect of dynamical noise in the CSW model. Twenty sample trajectories of ice volume $x(t)$ corresponding to different noise realizations with $\sigma = 0.002$

where $\langle x_i(t) \rangle = \frac{1}{N} \sum_{i=1}^{N} x_i(t)$ is the ensemble average of $N = 200$ trajectories. The size of dispersion $S(t)$ may be used as a measure of dynamical complexity induced by noise [40].

- *Number of large clusters at a time instant t, $N_{LC}(t)$.* For every time t, the N model states associated with the N sample solutions of the stochastic differential equation are grouped in clusters and elsewhere. The metric used for clustering the states of I80, CSW, SM90, and HA02 is simply the Euclidean distance in, respectively, the 1-, 2-, and 3-D spaces of climate variables **x**. Different approaches may be imagined for P98. One pragmatic and sufficiently robust solution is to define an auxiliary variable y taking values 1, 2, or 3 for states i, g, and G, respectively, and define the Euclidean distance in the (x, y) space. With this distance at hand, clusters are defined using the following iterative algorithm, similar to [19].

1. Define $\mathbb{I} = \{1, 2, \ldots, N\}$ the indices of the ensemble to be clustered, and call $\mathbf{x}_i(t)$ the ith model state.
2. Set $j = 1$ and repeat the following steps until $\mathbb{I}$ becomes empty:

   a. Call $\mathbf{x}^\star(t)$ the model state corresponding to one of the members of ensemble $\mathbb{I}$.
   b. Define $\mathbb{I}_j$, the ensemble of indices $\{i \in \mathbb{I} : ||\mathbf{x}^\star(t) - \mathbf{x}_i(t)|| < \epsilon\}$.
   c. Update $\mathbb{I} = \mathbb{I} \setminus \mathbb{I}_j$.
   d. Increment $j = j + 1$ if $\mathbb{I} \neq \emptyset$.

3. The clusters are the $\{\mathbb{I}_j\}$.

We use $\epsilon = 0.7$ for SM90 and HA02 and $\epsilon = 0.4$ for the other models. The number of *large* clusters, $N_{LC}(t)$, is defined as the number of clusters with at least ten members.

- *Finite-time Lyapunov exponent $\lambda_T(\mathbf{x}(t), \delta\mathbf{x}(t_0))$ for a time interval $[t, t + T]$* is defined as in [22]:

$$\lambda_T(\mathbf{x}(t), \delta\mathbf{x}(t_0)) = \frac{1}{T} \ln \frac{|\delta\mathbf{x}(t + T)|}{|\delta\mathbf{x}(t)|}, \tag{9}$$

where $\delta\mathbf{x}(t)$ is a vector representing an infinitesimal deviation from a reference trajectory of the climate state, $\mathbf{x}(t)$. The vector $\delta\mathbf{x}(t)$ is given as a solution of the linearized equation of the original dynamical system. The finite-time Lyapunov exponent $\lambda_T(\mathbf{x}(t), \delta\mathbf{x}(t_0))$ gives the rate of exponential divergence of nearby orbits from the reference trajectory $\mathbf{x}(t)$ during the time interval $[t, t+T]$. Typical initial deviations $\delta\mathbf{x}(t_0)$ give a same value for each trajectory $\mathbf{x}(t)$ for $t \gg t_0$. For simplicity, we denote $\lambda_T(\mathbf{x}(t), \delta\mathbf{x}(t_0))$ by $\lambda_T(t)$, but note that $\lambda_T(t)$ still depends on $\mathbf{x}(t)$. A positive (negative) value of $\lambda_T(t)$ indicates temporal instability (stability) of a trajectory in the time interval $[t, t + T]$. As $T \to \infty$, it converges to the largest Lyapunov exponent λ.

3 Results

3.1 Dispersions in Each Model

We compare the noise sensitivity of each model by using the maximum size of dispersion $S_{\max}$ and the mean size of dispersion S_{mean} during the time interval $[t_s, t_e]$:

$$S_{\max} = \max_{t_s \le t \le t_e} S(t),$$

$$S_{mean} = \frac{1}{t_e - t_s} \int_{t_s}^{t_e} S(t)dt.$$

First, we consider these quantities in a long time interval from $t_s = -700$ ka to $t_e = 100$ Ma, in order to characterise global properties of the attractor of each model. The maximum $S_{\max}$ and the mean S_{mean} are presented as functions of the scaled noise intensity $\sigma \in [0.001, 0.1]$ in Figs. 4a, b. The I80 model is fairly robust against dynamical noise, but the other models are highly sensitive to dynamical noise in the sense that large dispersions of trajectories, $S_{\max} \sim 1$, can be induced by extremely small noise (e.g. $\sigma \sim 0.001$) if one waits long enough (Fig. 4a). The mean size of dispersion S_{mean} is relatively large in SNA models (CSW and SM90) and the chaotic model (HA02), but it is small in the P98 model (Fig. 4b). These differences become less obvious for large noise $\sigma > 0.1$.

We now examine the qualitative differences between the dynamics of dispersions generated by each model in particular for small dynamical noise $\sigma = 0.002$. Figure 5a–e present the time series of the number of large clusters $N_{LC}(t)$ (top), the size of dispersion $S(t)$ (middle), and the finite-time Lyapunov exponent $\lambda_T(t)$ (bottom) for each model. No large dispersion appears in the I80 model. Large dispersions intermittently occur in the other nonchaotic systems (P98, CSW, and SM90) (cf. again Fig. 3). Periods of synchronisation ($N_{LC}(t) = 1$) may extend over several glacial cycles, unlike what is seen in the chaotic system (HA02) (though we note a period with two large clusters in the chaotic case, around $t = +5$ Ma).

Let us now focus on the intermittent dispersions. In the P98 model, the episodes of large dispersion are infrequent and relatively short (typically, one or two glacial cycles) because large dispersions may only occur when the system state is near one

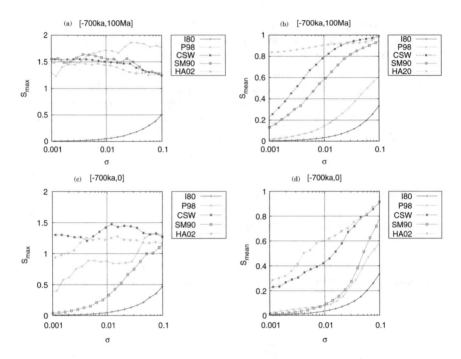

Fig. 4 Comparison of noise sensitivity between the models: (**a**) Maximal size of dispersion S_{max} in the long time interval $[-700\,ka, 100\,Ma]$. (**b**) Mean size of dispersion S_{mean} in the long time interval $[-700\,ka, 100\,Ma]$. (**c**) Maximal size of dispersion S_{max} in the time interval $[-700\,ka, 0]$. (**d**) Mean size of dispersion S_{mean} in the time interval $[-700\,ka, 0]$

of the discrete thresholds defined in the model. Elsewhere the system is stable. In the models with SNA (CSW and SM90), the episodes of large dispersion can last several million years in CSW and several tens of million years in SM90. These large dispersions are caused by temporal instability of the system. In fact, it may be observed that the original deterministic systems tend to have a large positive value of the finite-time Lyapunov exponent $\lambda_T(t)$ before the onsets of the dispersions (Figs. 5c, d (bottom)). The finite-time Lyapunov exponent of the system under the dynamical noise behaves similarly as Figs. 5c, d since the dynamical noise is small (data are not shown).

3.2   Order in Dispersions

In the models with SNAs (CSW and SM90), the long-lasting large dispersions are related to the existence of transient orbits with a long life time. The existence of transient orbits is illustrated in Fig. 6a (top, blue lines) using the CSW model: stochastic trajectories are generated with $\sigma = 0.002$ over the time interval $[-700\,ka, 0]$; then the noise is shut off and the system is integrated with the

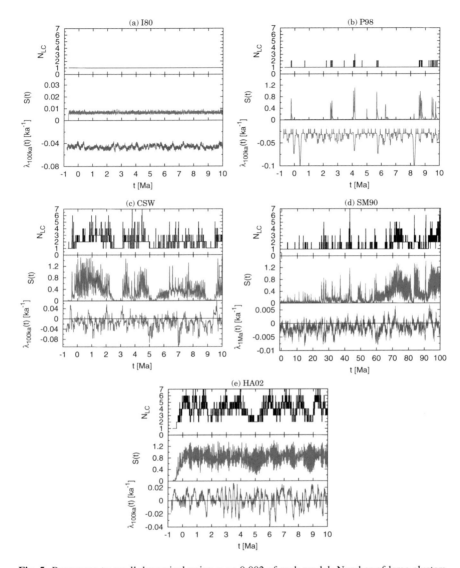

Fig. 5 Responses to small dynamical noise $\sigma = 0.002$ of each model: Number of large clusters $N_{LC}(t)$ (*top, black*). Size of dispersion $S(t)$ (*middle, red*). Finite-time Lyapunov exponent $\lambda_T(t)$ of the unperturbed system (*bottom, blue*). The averaging time T for $\lambda_T(t)$ is 1 Ma for SM90 and 100 ka for the other models. Note that the time interval of panel (**d**) SM90 is longer than the others because of the slow evolution of dispersion in this model

original deterministic equation. Transient orbits are excited by noise slightly before the deglaciation around $t = -400$ ka, where the finite-time Lyapunov exponent $\lambda_{200\,\mathrm{ka}}(t)$ is temporarily positive (Fig. 6a, bottom). These transient orbits may then last over more than 1 million years after the cessation of dynamical noise.

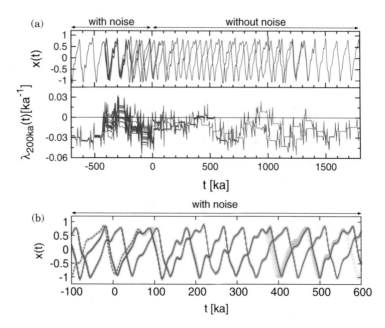

Fig. 6 Transient orbits with a long life time in CSW model: (**a**) The trajectory corresponding to the attractor of the original noiseless system (*red*) and some pieces of transient orbits with a long life time (*blue*) (*top*). The finite-time Lyapunov exponent $\lambda_{200\,ka}(t)$ for the trajectories in the *top panel* (*bottom*). (**b**) Twenty trajectories under the dynamical noise with $\sigma = 0.002$ (*green points*). The *red* and *blue lines* are the same in the *top panel* in (**a**)

They are attractive, in the sense that the finite-time Lyapunov exponent $\lambda_{200\,ka}(t)$ is negative on average in time (Fig. 6a, bottom, blue lines). As a result, when large dispersions occur, individual trajectories may get attracted either around the trajectory corresponding to the attractor of the original deterministic system or around some pieces of transient orbits that have a long life time, as shown in Fig. 6b.

The existence of such stable transient orbits was reported by Kapitaniak [39], where they were termed *strange nonchaotic transients*. They can also be related to the notion of finite-time attractivity, and more specifically that of (p, T)-attractor defined by Rasmussen [57] (pp. 19–20).

Dispersed trajectories in the models with SNA (CSW and SM90) are more clustered than those in the chaotic model (HA02). The number of large clusters $N_{LC}(t)$ in the models with SNA (CSW and SM90) is smaller on average than that in the chaotic model (HA02), as shown in Fig. 5c–e. Furthermore, the number of the points outside large clusters is larger in the chaotic model, as shown in Fig. 7. This result is intuitively reasonable: given that the largest Lyapunov exponent λ is positive but small in the chaotic model ($\lambda \approx 0.0020\,ka^{-1}$), we do expect trajectories to slowly travel away from the center of the clusters and thus distribute themselves over the phase space more widely than in the case of strange nonchaotic models.

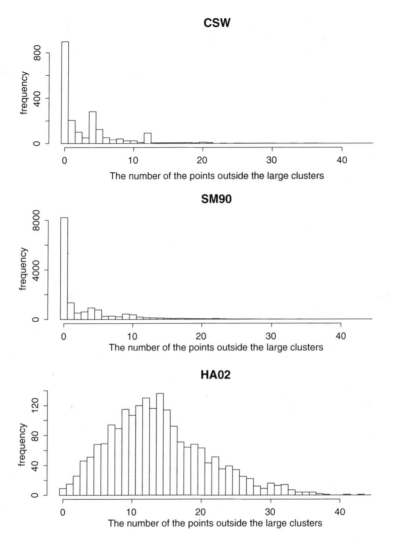

Fig. 7 Frequency distribution for the number of the points outside the large clusters. These results correspond to Fig. 5c–e

3.3 Implication for Ice Ages

The time interval focused so far, [−700 ka, 100 Ma], is quite long from the viewpoint of ice ages. Figure 4c, d present the maximum size S_{max} and the mean size S_{mean} of dispersions calculated in the "short" time interval [−700 ka, 0], where 100-ka glacial cycles took place in the history. In this short time window, the SM90 model with SNA is relatively robust against noise of $\sigma \sim O(0.01)$ because dispersions cannot evolve to the attractor size owing to its weak instability. On the other hand, the CSW model, also with SNA, is not robust against the noise of $\sigma \sim O(0.01)$ owing to its stronger temporal instability $\lambda_{100\,ka}(t) \sim 0.05\,ka^{-1}$ appeared around $t \sim -0.5\,Ma$ (Fig. 5c, bottom). To assess the stability of dynamical systems models of ice ages, it is useful to know not only the type of attractor (such as quasiperiodic, piecewise smooth, strange nonchaotic, or chaotic) but also the degree of the temporal instability, which may be characterised by the finite-time Lyapunov exponent $\lambda_T(t)$.

4 Concluding Discussion

This study shows the possibility that the sequence of glacial cycles can be temporally fragile in spite of the pacing by the astronomical forcing.

We studied the influence of dynamical noise in several models of glacial-interglacial cycles. This analysis outlines a form of temporarily quantised instability in systems that are characterised by an SNA (CSW and SM90). Specifically, the systems are synchronised on the astronomical forcing, but large dispersions of stochastic trajectories can be induced by extremely small noise at key times when the system is temporarily unstable. After a dispersion event, the trajectories are organised around a small number of clusters, which may co-exist over several glacial cycles until they merge again. The phenomenon is interpreted as a noise-induced excitation of long transient orbits. Dispersion events may be more or less frequent and, depending on the amount of noise, models with SNA may have very long horizons of predictability compared to the duration of geological periods.

Compared to this scenario, the model with a smooth attractor (I80) is always stable, i.e., large dispersion of orbits never occurs. On the other hand, the dynamics of the chaotic model (HA02) bare some similarities with the models with SNA, in the sense that trajectories cluster, i.e., at a given time the state of the system may confidently be located within a small number of regions. The difference is that there is a larger amount of leakage from the clusters, i.e., individual trajectories escape more easily from the cluster they belong to, and this reduces the predictability of such systems. Finally, we discussed a hybrid dynamical system with a piecewise continuous attractor (P98). Owing to the discontinuity of the attractor, very small amount of noise may rarely induce significant dispersion of trajectories, and contrarily to the scenario with SNA, there are no long transients because trajectories form a single cluster rapidly.

This analysis has some implications on the interpretation of statistics on the relationship between the timing of ice ages and astronomical forcing. In particular, the Rayleigh statistic was used to reject a null hypothesis of independence of the phase of glacial-interglacial cycles on the components of the astronomical forcing [34, 43]. Considering that even chaotic systems show the clustering of trajectories, our results show that Rayleigh statistics, alone, are not sufficient to determine whether the sequence of ice ages is stable or not.

In this article, we did not mention the notions in random dynamical systems theory [2, 60] so as to avoid a confusion between the classical, forward-type, definition of attractors (such as used for SNAs) and the pullback-type definition of attractors in random dynamical systems theory. However, it will be useful to formulate the present results in terms of random dynamical systems theory, where the noise-excited orbits around transient orbits may be reformulated as random fixed points. For example, in such a framework, the dynamical transition associated with a parameter change from stochastic SM90 and HA02 may be understood as a bifurcation from random fixed points to random strange attractors. Particular attention should then be paid to the nature of stochastic parameterisations and their effects on system stability.

Taking a wider prospective, this research along with other recent works on dynamical systems of ice ages [21, 59] may provide guidance for the design and interpretation of simulations with more sophisticated models. One of the targets of the palaeoclimate modelling community is to simulate ice ages by resolving the dynamics of the atmosphere, ocean, sea-ice and ice sheets, coupled with adequate representations of biogeochemical processes. Reasonable success has been achieved with atmosphere-ocean-ice-sheet-models forced by known CO_2 variation and astronomical forcing [1, 24, 26] but simulations of the fully coupled system have only appeared recently [25]. It would therefore be useful to determine the attractor properties associated with such complex numerical systems forced by the astronomical forcing, and in particular estimate to what extent they may generate long transients and display sensitive dependence to noise. The task involves both mathematical and technical challenges that will need to be addressed by steps of growing model complexity.

Appendix

The following sets of parameters are used in this study.

- I98 model [36]: $\tau = 17\,\text{ka}$, $b = 0.6$, $L = 0.249$, and $\epsilon = 0.4$.
- P98 model [51]: $X_i = 0$, $X_g = 1$, $X_G = 1$, $\tau_i = 10\,\text{ka}$, $\tau_g = 50\,\text{ka}$, $\tau_G = 50\,\text{ka}$, $\tau_F = 25\,\text{ka}$, $i_0 = -0.75$, $i_1 = 0$, $a = 1$, $L = 0.307$, and $\epsilon = 0.4$.
- CSW model [16]: $\alpha = 30$, $\beta = 0.75$, $\gamma = 0.4$, $\tau = 36\,\text{ka}$, $L = 0.810$, and $\epsilon = 0.4$.
- SM90 model [63]: $p = 1$, $q = 2.5$, $r = 0.9$, $s = 1$, $u = 0.6$, $v = 0.2$, $\tau = 10\,\text{ka}$, $L = 0.546$, and $\epsilon = 0.7$.
- HA02 model [30]: $p = 0.82$, $q = 2.5$, $r = 0.95$, $s = 0.53$, $u = 0.32$, $v = 0.02$, $\tau = 10\,\text{ka}$, $L = 0.726$, and $\epsilon = 0.7$.

Acknowledgements MC is senior research associate with the Belgian National Fund of Scientific Research. This research is a contribution to the ITOP project, ERC-StG 239604 and to the Belgian Federal Policy Office project BR/121/A2/STOCHCLIM.

References

1. Abe-Ouchi, A., Saito, F., Kawamura, K., Raymo, M.E., Okuno, J., Takahashi, K., Blatter, H.: Insolation-driven 100,000-year glacial cycles and hysteresis of ice-sheet volume. Nature **500**(7461), 190–193 (2013). doi:10.1038/nature12374
2. Arnold, L.: Random Dynamical Systems. Springer Monographs in Mathematics. Springer, Berlin/Heidelberg (1998) doi:10.1007/978-3-662-12878-7
3. Ashkenazy, Y.: The role of phase locking in a simple model for glacial dynamics. Clim. Dyn. **27**, 421–431 (2006). doi:10.1007/s00382-006-0145-5
4. Benzi, R., Parisi, G., Sutera, A., Vulpiani, A.: Stochastic resonance in climatic change. Tellus **34**(1), 10–16 (1982). doi:10.1111/j.2153-3490.1982.tb01787.x
5. Berger, A.: Support for the astronomical theory of climatic change. Nature **268**, 44–45 (1977). doi:10.1038/269044a0
6. Berger, A.: Milankovitch theory and climate. Rev. Geophys. **26**(4), 624–657 (1988)
7. Berger, A., Loutre, M.: Insolation values for the climate of the last 10 million years. Quat. Sci. Rev. **10**(4), 297–317 (1991). doi:10.1016/0277-3791(91)90033-Q
8. Berger, A., Loutre, M.F.: Origine des fréquences des éléments astronomiques intervenant dans l'insolation. Bull. Classe des Sci. **1–3**, 45–106 (1990)
9. Berger, A., Loutre, M.F.: Precession, eccentricity, obliquity, insolation and paleoclimates. In: Duplessy, J.C., Spyridakis, M.T. (eds.) Long-Term Climatic Variations. NATO ASI Series, vol. 122, pp. 107–151. Springer, Heidelberg (1994)
10. Berger, A., Melice, J.L., Loutre, M.F.: On the origin of the 100-kyr cycles in the astronomical forcing. Paleoceanography **20**(PA4019) (2005). doi:10.1029/2005PA001173
11. Berger, A.L.: Long-term variations of daily insolation and Quaternary climatic changes. J. Atmos. Sci. **35**, 2362–2367 (1978). doi:10.1175/1520-0469(1978)035<2362:LTVODI>2.0.CO;2
12. Bolton, E.W., Maasch, K.A., Lilly, J.M.: A wavelet analysis of plio-pleistocene climate indicators – a new view of periodicity evolution. Geophys. Res. Lett. **22**(20), 2753–2756 (1995). doi:10.1029/95GL02799
13. Broecker, W.S., van Donk, J.: Insolation changes, ice volumes and the O^{18} record in deep-sea cores. Rev. Geophys. **8**(1), 169–198 (1970). doi:10.1029/RG008i001p00169
14. Calov, R., Ganopolski, A., Claussen, M., Petoukhov, V., Greve, R.: Transient simulation of the last glacial inception. Part I: glacial inception as a bifurcation in the climate system. Clim. Dyn. **24**(6), 545–561 (2005). doi:10.1007/s00382-005-0007-6
15. Crucifix, M.: How can a glacial inception be predicted? The Holocene **21**(5), 831–842 (2011). doi:10.1177/0959683610394883
16. Crucifix, M.: Oscillators and relaxation phenomena in pleistocene climate theory. Philos. Trans. R. Soc. A Math. Phys. Eng. Sci. **370**(1962), 1140–1165 (2012). doi:10.1098/rsta.2011.0315. http://rsta.royalsocietypublishing.org/content/370/1962/1140. abstract
17. Crucifix, M.: Why could ice ages be unpredictable? Clim. Past **9**(5), 2253–2267 (2013).doi:10.5194/cp-9-2253-2013
18. Daruka, I., Ditlevsen, P.: A conceptual model for glacial cycles and the middle pleistocene transition. Clim. Dyn. 1–12 (2015). http://dx.doi.org/10.1007/s00382-015-2564-7
19. De Saedeleer, B., Crucifix, M., Wieczorek, S.: Is the astronomical forcing a reliable and unique pacemaker for climate? A conceptual model study. Clim. Dyn. **40**, 273–294 (2013). doi:10.1007/s00382-012-1316-1
20. Ditlevsen, P.: Observation of α-stable noise induced millennial climate changes from an ice-core record. Geophys. Res. Lett. **26**(10), 1441–1444 (1999). doi:10.1029/1999GL900252
21. Ditlevsen, P.D.: Bifurcation structure and noise-assisted transitions in the Pleistocene glacial cycles. Paleoceanography **24**, PA3204 (2009). doi:10.1029/2008PA001673
22. Eckhardt, B., Yao, D.: Local Lyapunov exponents in chaotic systems. Phys. D Nonlinear Phenom. **65**(1–2), 100–108 (1993). doi:10.1016/0167-2789(93)90007-N

23. Feudel, U., Kuznetsov, S., Pikovsky, A.: Strange Nonchaotic Attractors: Dynamics Between Order And Chaos in Quasiperiodically Forced Systems. World Scientific Series on Nonlinear Science, Series A Series. World Scientific (2006). http://books.google.be/books?id=zptIMMkI2kEC

24. Gallée, H., van Ypersele, J.P., Fichefet, T., Marsiat, I., Tricot, C., Berger, A.: Simulation of the last glacial cycle by a coupled, sectorially averaged climate-ice sheet model. Part II: response to insolation and CO_2 variation. J. Geophys. Res. **97**, 15713–15740 (1992). doi:10.1029/92JD01256

25. Ganopolski, A., Brovkin, V., Calov, R.: Robustness of quaternary glacial cycles. Geophys. Res. Abstr. **17**, EGU2015–7197–1 (2015)

26. Ganopolski, A., Calov, R.: Simulation of glacial cycles with an earth system model. In: Berger, A., Mesinger, F., Sijacki, D. (eds.) Climate Change: Inferences from Paleoclimate and Regional Aspects, pp. 49–55. Springer, Vienna (2012). doi:10.1007/978-3-7091-0973-1_3

27. Gildor, H., Tziperman, E.: A sea ice climate switch mechanism for the 100-kyr glacial cycles. J. Geophys. Res. (Oceans) **106**, 9117–9133 (2001)

28. Grebogi, C., Ott, E., Pelikan, S., Yorke, J.A.: Strange attractors that are not chaotic. Phys. D Nonlinear Phenom. **13**(1–2), 261–268 (1984). doi:10.1016/0167-2789(84)90282-3

29. Greiner, A., Strittmatter, W., Honerkamp, J.: Numerical integration of stochastic differential equations. J. Stat. Phys. **51**(1–2), 95–108 (1988). doi:10.1007/BF01015322

30. Hargreaves, J.C., Annan, J.D.: Assimilation of paleo-data in a simple Earth system model. Clim. Dyn. **19**, 371–381 (2002). doi:10.1007/s00382-002-0241-0

31. Hasselmann, K.: Stochastic climate models part I. Theory. Tellus **28**(6), 473–485 (1976). doi:10.1111/j.2153-3490.1976.tb00696.x

32. Hays, J.D., Imbrie, J., Shackleton, N.J.: Variations in the Earth's orbit: pacemaker of ice ages. Science **194**, 1121–1132 (1976). doi:10.1126/science.194.4270.1121

33. Huybers, P.: Pleistocene glacial variability as a chaotic response to obliquity forcing. Clim. Past **5**(3), 481–488 (2009). doi:10.5194/cp-5-481-2009

34. Huybers, P., Wunsch, C.: Obliquity pacing of the late Pleistocene glacial terminations. Nature **434**, 491–494 (2005). doi:10.1038/nature03401

35. Imbrie, J., Berger, A., Boyle, E.A., Clemens, S.C., Duffy, A., Howard, W.R., Kukla, G., Kutzbach, J., Martinson, D.G., McIntyre, A., Mix, A.C., Molfino, B., Morley, J.J., Peterson, L.C., Pisias, N.G., Prell, W.L., Raymo, M.E., Shackleton, N.J., Toggweiler, J.R.: On the structure and origin of major glaciation cycles. Part 2: the 100, 000-year cycle. Paleoceanography **8**, 699–735 (1993). doi:10.1029/93PA02751

36. Imbrie, J., Imbrie, J.Z.: Modelling the climatic response to orbital variations. Science **207**, 943–953 (1980). doi:10.1126/science.207.4434.943

37. Kanamaru, T.: Duffing oscillator. Scholarpedia **3**(3), 6327 (2008). Revision #91210

38. Kaneko, K.: Oscillation and doubling of torus. Prog. Theor. Phys. **72**(2), 202–215 (1984). doi:10.1143/PTP.72.202. http://ptp.oxfordjournals.org/content/72/2/202.abstract

39. Kapitaniak, T.: Strange non-chaotic transients. J. Sound Vib. **158**(1), 189–194 (1992). doi:10.1016/0022-460X(92)90674-M

40. Khovanov, I.A., Khovanova, N.A., McClintock, P.V.E., Anishchenko, V.S.: The effect of noise on strange nonchaotic attractors. Phys. Lett. A **268**(4–6), 315–322 (2000). doi:10.1016/S0375-9601(00)00183-3. http://www.sciencedirect.com/science/article/pii/S0375960100001833

41. Lambeck, K., Chapell, J.: Sea level change throughout the last glacial cycle. Science **292**, 679–686 (2001). doi:10.1126/science.1059549

42. Le Treut, H., Ghil, M.: Orbital forcing, climatic interactions and glaciation cycles. J. Geophys. Res. **88**(C9), 5167–5190 (1983). doi:10.1029/JC088iC09p05167

43. Lisiecki, L.E.: Links between eccentricity forcing and the 100,000-year glacial cycle. Nat. Geosci. **3**(5), 349–352 (2010). doi:10.1038/ngeo828

44. Lisiecki, L.E., Raymo, M.E.: A Pliocene-Pleistocene stack of 57 globally distributed benthic $\delta^{18}O$ records. Paleoceanography **20**, PA1003 (2005). doi:10.1029/2004PA001071

45. Milankovitch, M.: Canon of insolation and the ice-age problem. Narodna biblioteka Srbije, Beograd (1998). English translation of the original 1941 publication

46. Mitsui, T., Aihara, K.: Dynamics between order and chaos in conceptual models of glacial cycles. Clim. Dyn. **42**(11–12), 3087–3099 (2014). doi:10.1007/s00382-013-1793-x
47. Mitsui, T., Crucifix, M., Aihara, K.: Bifurcations and strange nonchaotic attractors in a phase oscillator model of glacial–interglacial cycles. Phys. D Nonlinear Phenom. **306**, 25–33 (2015). doi:10.1016/j.physd.2015.05.007
48. Nicolis, C.: Stochastic aspects of climatic transitions—response to a periodic forcing. Tellus **34**(1), 1–9 (1982). doi:10.1111/j.2153-3490.1982.tb01786.x. http://dx.doi.org/10.1111/j.2153-3490.1982.tb01786.x
49. Nicolis, C.: Climate predictability and dynamical systems. In: Nicolis, C., Nicolis, G. (eds.) Irreversible Phenomena and Dynamical System Analysis in the Geosciences. NATO ASI series C: Mathematical and Physical Sciences, vol. 192, pp. 321–354. Kluwer, Dordrecht (1987)
50. Nishikawa, T., Kaneko, K.: Fractalization of a torus as a strange nonchaotic attractor. Phys. Rev. E **54**, 6114–6124 (1996). doi:10.1103/PhysRevE.54.6114
51. Paillard, D.: The timing of Pleistocene glaciations from a simple multiple-state climate model. Nature **391**, 378–381 (1998). doi:10.1038/34891
52. Paillard, D., Parrenin, F.: The Antarctic ice sheet and the triggering of deglaciations. Earth Planet. Sci. Lett. **227**, 263–271 (2004). doi:10.1016/j.epsl.2004.08.023
53. Pelletier, J.D.: Coherence resonance and ice ages. J. Geophys. Res. Atmos. **108**(D20), (2003). doi:10.1029/2002JD003120. http://dx.doi.org/10.1029/2002JD003120
54. Penland, C.: Noise out of chaos and why it won't go away. Bull. Am. Meteorol. Soc. **84**(7), 921–925 (2003). doi:10.1175/BAMS-84-7-921. http://journals.ametsoc.org/doi/abs/10.1175/BAMS-84-7-921
55. Petit, J.R., Jouzel, J., Raynaud, D., Barkov, N.I., Barnola, J.M., Basile, I., Bender, M., Chappellaz, J., Davis, M., Delaygue, G., Delmotte, M., Kotlyakov, V.M., Legrand, M., Lipenkov, V.Y., Lorius, C., Pepin, L., Ritz, C., Saltzman, E., Stievenard, M.: Climate and atmospheric history of the past 420, 000 years from the Vostok ice core, Antarctica. Nature **399**, 429–436 (1999). doi:10.1038/20859
56. Prasad, A., Negi, S.S., Ramaswamy, R.: Strange nonchaotic attractors. Int. J. Bifurc. Chaos **11**(2), 291–309 (2001). doi:10.1142/S0218127401002195
57. Rasmussen, M.: Attractivity and Bifurcation for Nonautonomous Dynamical Systems. No. 1907 in Lecture Notes in Mathematics. Springer, Berlin/Heidelberg (2000)
58. Raymo, M.: The timing of major climate terminations. Paleoceanography **12**(4), 577–585 (1997). doi:10.1029/97PA01169
59. Rial, J.A., Oh, J., Reischmann, E.: Synchronization of the climate system to eccentricity forcing and the 100,000-year problem. Nat. Geosci. **6**(4), 289–293 (2013). doi:10.1038/ngeo1756
60. Roques, L., Chekroun, M.D., Cristofol, M., Soubeyrand, S., Ghil, M.: Parameter estimation for energy balance models with memory. Proc. R. Soc. A **470**, 20140349 (2014). doi:10.1098/rspa.2014.0349
61. Ruddiman, W.F.: Orbital changes and climate. Q. Sci. Rev. **25**, 3092–3112 (2006). doi:10.1016/j.quascirev.2006.09.001
62. Saltzman, B., Maasch, K.A.: Carbon cycle instability as a cause of the late Pleistocene ice age oscillations: modeling the asymmetric response. Glob. Biogeochem. Cycles **2**(2), 117–185 (1988). doi:10.1029/GB002i002p00177
63. Saltzman, B., Maasch, K.A.: A first-order global model of late Cenozoic climate. Trans. R. Soc. Edinb. Earth Sci. **81**, 315–325 (1990). doi:10.1017/S0263593300020824
64. Shackleton, N.J., Opdyke, N.D.: Oxygen isotope and paleomagnetic stratigraphy of equatorial Pacific core V28-239: late Pliocene to latest Pleistocene. In: Cline, R.M., Hays, J.D. (eds.) Investigation of Late Quaternary Paleoceanography and Paleoclimatology. Memoir, vol. 145, pp. 39–55. Geological Society of America, Boulder (1976)
65. Sturman, R., Stark, J.: Semi-uniform ergodic theorems and applications to forced systems. Nonlinearity **13**(1), 113 (2000). http://stacks.iop.org/0951-7715/13/i=1/a=306
66. Thompson, D.J.: Pectrum estimation and harmonic analysis. Proc. IEEE **70**, 1055–1096 (1982)
67. Timmermann, A., Lohmann, G.: Noise-induced transitions in a simplified model of the thermohaline circulation. J. Phys. Oceanogr. **30**, 1891–1900 (2000)

68. Tziperman, E., Raymo, M.E., Huybers, P., Wunsch, C.: Consequences of pacing the Pleistocene 100 kyr ice ages by nonlinear phase locking to Milankovitch forcing. Paleoceanography **21**, PA4206 (2006). doi:10.1029/2005PA001241
69. Vautard, R., Yiou, P., Ghil, M.: Singular-spectrum analysis: a toolkit for short, noisy chaotic signals. Phys. D Nonlinear Phenom. **58**(1–4), 95–126 (1992). doi:10.1016/0167-2789(92)90103-T
70. Weertman, J.: Milankovitch solar radiation variations and ice age ice sheet sizes. Nature **261**, 17–20 (1976). doi:10.1038/261017a0
71. Wunsch, C.: The spectral description of climate change including the 100ky energy. Clim. Dyn. **20**, 353–363 (2003). doi:10.1007:S00382-002-0279-z

On the Calibration of Lévy Driven Time Series with Coupling Distances and an Application in Paleoclimate

Jan Gairing, Michael Högele, Tetiana Kosenkova, and Alexei Kulik

Abstract This article aims at the statistical assessment of time series with large fluctuations in short time, which are assumed to stem from a continuous process perturbed by a Lévy process exhibiting a heavy tail behavior. We propose an easily implementable procedure to estimate efficiently the statistical difference between the noise process generating the data and a given reference jump measure in terms of so-called *coupling distances* introduced in Gairing et al. (Stoch Dyn 15(2):1550009-1–1550009-25, 2014). After a short introduction to Lévy processes and coupling distances we recall basic statistical approximation results and derive asymptotic rates of convergence. In the sequel the procedure is elaborated in detail in an abstract setting and eventually applied in a case study with simulated and paleoclimate proxy data. Our statistic indicates the dominant presence of a non-stable heavy-tailed jump Lévy component for some tail index $\alpha > 2$ in the paleoclimatic record.

Keywords Time series with heavy tails • Index of stability • Goodness-of-fit • Empirical Wasserstein distance • Limit theorems • Empirical quantile process

MSC 2010: 60G51; 60G52; 60J75; 62M10; 62P12

J. Gairing
Institut für Mathematik, Humboldt-Universität zu Berlin, Unter den Linden 6, 10099 Berlin, Germany
e-mail: gairing@math.hu-berlin.de

M. Högele (✉)
Departamento de Matemáticas, Universidad de los Andes, Cra 1 No 18 A, Bogotá, Colombia
e-mail: ma.hoegele@uniandes.edu.co

T. Kosenkova
Institut für Mathematik, Universität Potsdam, Karl-Liebknecht-Str. 24-25, 14476 Potsdam, Germany
e-mail: kosenkova@math.uni-potsdam.de

A. Kulik
Institute of Mathematics, Ukrainian National Academy of Sciences, 3 Tereshchenkivska, 01601 Kyiv, Ukraine
e-mail: kulik.alex.m@gmail.com

© Springer International Publishing Switzerland 2016
F. Ancona et al. (eds.), *Mathematical Paradigms of Climate Science*, Springer
INdAM Series 15, DOI 10.1007/978-3-319-39092-5_7

1 Introduction

In many contexts the fluctuations found in time series of interest exceed the level which is plausible for an underlying continuous model. The continuity of Gaussian models makes it necessary to go beyond the Gaussian paradigm to model random perturbations and to include the effect of shocks. The natural class including discontinuous perturbations is given by Lévy processes, that is non-Gaussian extensions of Brownian motion, which keep the white noise structure of stationary, δ-correlated increments. Often stochastic modeling consists in the study of deterministic models, which represent preknowledge about the underlying phenomenon, perturbed by (Lévy) noise.

In this contribution we want to follow the ideas of [9] in order to calibrate the jump behavior in our model by means of *coupling distances* based on empirical data. Coupling distances contain a suitably renormalized Wasserstein distance between jump measures of Lévy processes, which metrizes the weak convergence of distributions. In their work they consider a class of Lévy driven dynamical systems and derive a quantitative upper bound for the proximity of their distributions on sample path space. An essential factor in this estimate is the coupling distance between Lévy measures. By construction these distances explore the discrepancy of Lévy measures along decreasing jump sizes. This article wishes to provide the theoretical background as well as an instructive road map to implement the concept of coupling distances in a statistical setting.

The estimation of the tail behavior is also of interest on its own. Many important dynamical features, stability properties or scaling invariance are determined by the mass distribution in the tails. See for instance [11, 14–16, 19]. In a mathematical paradigm of climate science rapid transitions between stadials and interstadials of the last glaciation period can be described by the impact of unpredictable shocks, that can be interpreted as large discontinuities of a random driving force. In the seminal article [6] Ditlevsen identified an α-stable Lévy component in a climate proxy signal with the help of the statistical analysis of large jumps. Further investigation in that direction has been carried out in [10] and [12] exploiting the selfsimilarity of these processes.

We will concentrate on the calibration of the tails of compound Poisson processes, a crucial subclass of Lévy processes with jumps. We refer to Sect. 2 for a nutshell and to [1] and [21] for the interested reader. This section also contains the definition of the coupling distance. In Sect. 3 we provide the statistical tools to assess the Wasserstein component in the coupling distance. We an asymptotic rate of convergence. Section 4 details the conceptual and algorithmic approach to the calibration problem. Conceptually we describe the modeling class interpreting large increments as governed by large jumps of the driving noise. This allows to define the statistical quantity of interest, the empirical version of a coupling distance. We present an algorithm how to evaluate this statistic on a data sample. In Sect. 5 we exemplify the previous strategy for the calibration of polynomial tails. Simulated sample paths confirm the strength of the method. Eventually we reassess the climate proxies studied by Ditlevsen et al. with our calibration procedure. Comparing the coupling distance for $\alpha = 1.75$ with values $\alpha > 2$ justifies the assumption of an non-stable jump Lévy component.

2 Lévy Processes and Coupling Distances

2.1 Lévy Processes

The most prominent representative of a Lévy process is arguably Brownian motion. A standard Brownian motion $B = (B_t)_{t \geqslant 0}$ is defined as a stochastic process starting in $B_0 = 0$, with independent increments satisfying the stationarity condition $B_t - B_s = B_{t-s} \sim N(0, (t - s))$ for $t \geqslant s$. It has almost surely continuous trajectories $t \mapsto B_t$ and moments of all orders.

The concept of a Lévy process drops the assumption of the Gaussianity of increments. A Lévy process is given as a stochastic process with independent and stationary increments $L_t - L_s \sim L_{t-s}$. Such a process is not necessarily continuous. However, to ensure the separability of the process (trajectories should be determined by any countable dense set of points in time) one imposes stochastic continuity, i.e.

$$\lim_{t \to s} \mathbb{P}(|L_t - L_s| > \eta) = 0 \quad \text{for all } \eta > 0.$$

This property ensures that almost all trajectories are at least right-continuous and with finite left limits (commonly denoted by the French acronym càdlàg). The so-called Lévy-Itô decomposition tells us that a given Lévy process $L = (L_t)_{t \geqslant 0}$ in $\mathbb{R}^d$ can be decomposed almost surely into the sum of four independent components, three of which are stochastic processes. That is there exits a vector $a \in \mathbb{R}^d$, a d-dimensional standard Brownian motion $B = (B_t)_{t \geqslant 0}$, a positive semi-definite (covariance) matrix $A \in \mathbb{R}^{d \otimes d}$ and a measure on the (Borel) measurable sets $\nu : \mathcal{M}(\mathbb{R}^d) \to [0, \infty]$ such that for all $\rho > 0$

$$\nu(\{0\}) = 0, \qquad \int_{\mathcal{B}_\rho(0)} |x|^2 \nu(dx) < \infty \qquad \text{and} \qquad \nu(\mathbb{R}^d \setminus \mathcal{B}_\rho(0)) < \infty, \quad (1)$$

where $\mathcal{B}_\rho(0) = \{y \in \mathbb{R}^d \mid |y| < \rho\}$. For each fixed $\rho > 0$ the measure ν has two associated processes. First, there is a compound Poisson process $C^\rho = (C_t^\rho)_{t \geqslant 0}$ with intensity $\lambda_\rho = \nu(\mathbb{R}^d \setminus \mathcal{B}_\rho(0))$ and jump distribution ν_ρ

$$\mathcal{M}(\mathbb{R}^d) \ni E \mapsto \nu_\rho(E) := \frac{\nu(E \cap (\mathbb{R}^d \setminus \mathcal{B}_\rho(0)))}{\lambda_\rho}. \quad (2)$$

That is a pure jump process with exponentially distributed waiting times between consecutive jumps of intensity λ_ρ, which are independent and distributed according to ν_ρ in (2). Note that the jumps of C^ρ are bounded from below by ρ. Second, there is another pure jump process $J^\rho = (J_t^\rho)_{t \geqslant 0}$, whose jumps are bounded from above by ρ. The Lévy process L is decomposed path-wise as

$$L_t = at + A^{1/2}B_t + C_t^\rho + J_t^\rho \qquad \forall t \geqslant 0 \quad \mathbb{P} - \text{a.s.} \qquad (3)$$

In particular, the marginal laws of L are given by the so-called Lévy-Chinchine representation of the characteristic function

$$\mathbb{E}[\exp(i\langle u, L_t\rangle)] = \exp(t\Psi(u)), \qquad t \geqslant 0, u \in \mathbb{R}^d,$$

with

$$\Psi(u) = iat - \frac{1}{2}\langle a, Aa\rangle + \int_{\mathbb{R}^d \setminus \mathscr{B}_\rho(0)} \left[e^{i\langle u,y\rangle} - 1\right]v(dy)$$

$$+ \int_{\mathscr{B}_\rho(0)} \left[e^{i\langle u,y\rangle} - 1 - i\langle u, y\rangle\right]v(dy), \qquad u \in \mathbb{R}^d.$$

This representation tells us that $J^\rho = (J_t^\rho)_{t\geqslant 0}$ can be understood as the superposition of independent (recentered) compound Poisson processes, with jumps that take values in rings given by $R_j = \{\rho_j < |y| \leqslant \rho_{j-1}\}$, for a strictly decreasing sequence $\rho = \rho_0 > \rho_1 > \cdots > 0$ of radii $\rho_j \searrow 0$, such that

$$\mathscr{B}_\rho(0) \setminus \{0\} = \bigcup_{j\in\mathbb{N}_0} R_j$$

with joint (possibly infinite) intensity

$$\sum_{j\in\mathbb{N}_0} v(R_j) = v(\mathscr{B}_\rho(0)) \leqslant \infty$$

and individual jump distribution $E \mapsto v_{\rho_{j+1}}(E) - v_{\rho_j}(E)$. If $v(\mathscr{B}_\rho(0)) < \infty$ we can choose formally $\rho = 0$ and hence $J^\rho \equiv 0$.

Remark 1 This construction relies heavily on the fact that the integrals in the exponent Ψ are additive over disjoint supports. Hence the characteristic function decomposes into a product of characteristic functions and thus independent components. As a consequence for every repartition $V_1, \ldots, V_n$ of $\mathbb{R}^d$ ($\bigcup_i V_i = \mathbb{R}^d$ and $V_i \cap V_j = \emptyset$ unless $i = j$) the Lévy process decomposes into the sum of n independent Lévy processes L^i with jumps taking values in V_i.

2.2   *Coupling Distances Between Lévy Measures*

In the article [9] the authors construct a distance on the set of Lévy measures in $\mathbb{R}^d$, that exploits the approximation of the discontinuous part of L by compound Poisson processes with decreasing lower bound on the jump size.

The idea is that given two jump (probability) distributions μ and μ' on $\mathbb{R}^d$ we want to measure the distance $\mathbb{E}[|X - Y|^2]$ between two random variables $X \sim \mu$ and $Y \sim \mu'$ on a common probability space. Obviously, there is more than one distribution of the random vector (X, Y), that guarantees the marginals $X \sim \mu$ and $Y \sim \mu'$. Any probability measure Π on the product space $\mathbb{R}^d \times \mathbb{R}^d$ with these marginals, is called a coupling of μ and μ'. Hence the set of all couplings is given by

$$\mathscr{C}(\mu, \mu') := \left\{ \Pi : \mathscr{M}(\mathbb{R}^d) \otimes \mathscr{M}(\mathbb{R}^d) \to [0, 1] \text{ probability measure, with} \right.$$

$$\left. \Pi(E \times \mathbb{R}^d) = \mu(E), \quad \Pi(\mathbb{R}^d \times E) = \mu'(E) \quad \text{for all } E \in \mathscr{M}(\mathbb{R}^d) \right\}.$$
(4)

Since the expectation operator is a functional entirely determined by $\Pi \in \mathscr{C}(\mu, \mu')$, the proximity of μ and μ' can be quantified minimizing the function $\Pi \mapsto \mathbb{E}_{\Pi}[|X - Y|^2]$ over the set of all couplings in $\mathscr{C}(\mu, \mu')$. This motivates the following abstract definition.

Definition 1 The Wasserstein metric of order 2 between two probability measures μ, μ' on $(\mathbb{R}^d, |\cdot|)$ is defined by

$$W_2(\mu, \mu') := \inf_{\substack{(X,Y) \sim \Pi \\ \Pi \in \mathscr{C}(\mu, \mu')}} \mathbb{E}[|X - Y|^2]^{\frac{1}{2}}.$$
(5)

Any minimizer is referred to as optimal coupling between μ and μ'.

Definition (5) should be understood as follows. We take the minimum of the L^2 distance between X and Y, ranging over all joint laws Π of random vectors $(X, Y) \sim \Pi$ such that $X \sim \mu$ and $Y \sim \mu'$. This is non-trivial since X and Y are not independent and there there are many joint laws Π in the product space, reflecting all kinds of dependencies of X and Y. It is well-known in the mathematical literature [20] that the convergence $W_2(\mu_n, \mu) \to 0$ is equivalent to the weak convergence $\mu_n \rightharpoonup \mu$ and the convergence of the second moments. We should stress the dependence of W_2 on the metric on the state space $(\mathbb{R}^d, |\cdot|)$ appearing under the expectation.

Remark 2 For instance on $\mathbb{R}$ with the Euclidean distance $|\cdot|$ one can calculate the optimal coupling explicitly. It can be shown that for two distribution functions $F(x) = \mu((-\infty, x])$ and $F'(x) = \mu'((-\infty, x])$ the optimal coupling is realized by the random vector

$$(X, Y) = (F^{-1}(U), (F')^{-1}(U)),$$

where $F^{-1}(u) = \inf\{x \in \mathbb{R} : F(x) \geq u\}$ is the *quantile function* or right inverse of F and U has the uniform distribution in $[0, 1]$. The same definition applies for F'.

Therefore the Wasserstein metric is easily evaluated by

$$W_2^2(\mu, \mu') = \int_0^1 |(F^{-1}(u) - (F')^{-1}(u)|^2 du.$$

The optimality of the pair (X, Y) relies on the specific metric on $\mathbb{R}$. The law of (X, Y) is obviously a coupling of μ and μ', and the right-hand side provides at least an upper bound for the Wasserstein distance.

In this work, however, we shall consider the following cutoff distance in order to account for random variables without second moments.

Example 1 Consider now $\mathbb{R}$ equipped with the cutoff absolute value $|\cdot|_1$ with $|x|_1 = \min(|x|, 1)$ and the Euclidean norm $|\cdot|$. This will be the right space to approximate the laws of the jumps. We introduce the cutoff norm, since we do not require the jumps to have second moments. Note in particular that the optimal coupling with respect to $|\cdot|_1$ is not known in general. That is $(X, Y) = (F^{-1}(U), (F')^{-1}(U))$ for U uniformly distributed in $[0, 1]$ still is a nice coupling since it can be calculated explicitly in Sect. 3, however not optimal with respect to $|\cdot|_1$. Instead in Sect. 3 it will serve as an upper bound of the Wasserstein distance with respect to $|\cdot|_1$.

Definition 2 For two absolutely continuous Lévy measures $v = f dx$ and $v' = f' dx$ on $\mathbb{R}^d$ and $0 < \lambda < \min(v(\mathbb{R}^d), v'(\mathbb{R}^d))$ let

$$\rho(\lambda) := \inf\{r > 0 \mid v(\mathbb{R}^d \setminus \mathscr{B}_r(0)) \geq \lambda\}$$

and $\rho'(\lambda)$ analogously. We introduce a family of semimetrics T_λ

$$T_\lambda(v, v') := \lambda^{\frac{1}{2}} W_2(v_{\rho(\lambda)}, v'_{\rho'(\lambda)}).$$

Remark 3 First note that given v and $\lambda > 0$ we have the following equality $\lambda_{\rho(\lambda)} = \lambda$, with λ_ρ defined before (2). Intuitively, given Lévy measures μ and μ', the semimetric $T_\lambda(v, v')$ compares the jump distributions $v_{\rho(\lambda)}$ and $v'_{\rho'(\lambda)}$ of two compound Poisson processes with common rate $\lambda = \lambda_{\rho(\lambda)} = \lambda_{\rho'(\lambda)}$.

Remark 4 Note further that the bivariate function T_λ is symmetric and satisfies the triangle inequality. Hence it is a semimetric. Clearly $T_\lambda(v, v) = 0$ does not guarantee that v is the zero measure, since $v|_{B_{\rho(\lambda)}}$ is not taken into account. Therefore it is not a proper metric. In order to overcome this shortcoming and provided that $v(\mathbb{R}^d) = v'(\mathbb{R}^d) = \infty$ we introduce the following.

Definition 3 For two absolutely continuous Lévy measures $v = f dx$ and $v' = f' dx$ on $\mathbb{R}^d$ with $v(\mathbb{R}^d) = v'(\mathbb{R}^d) = \infty$ we define

$$T(v, v') := \sup_{\lambda > 0} T_\lambda(v, v').$$

Remark 5 Both restrictions in the definitions above can be removed. For details we refer to the original work [9].

- The restriction on absolute continuity is overcome by an interpolation procedure.
- The requirement of infinite mass can be dropped by the ad hoc introduction of an artificial point mass in 0 carrying the missing weight. In this way also finite Lévy measures fall into the setup.

We also refer to [18] for further applications of coupling distances.

3   Statistical Considerations

In this section we provide the technical background to compare the jump statistics of a data set to a given reference distribution in terms of the coupling distance. For this purpose we collect the necessary statistical theory. Since our case study is essentially one dimensional we stick to the scalar case. For higher dimensions we refer to Remark 6 at the end of this section.

3.1   Basic Notions

Let us consider a sequence of independent and identically distributed random variables $(X_i)_{i \in \mathbb{N}}$ with common law μ on the real line. Denote by μ_n the *empirical distribution* based on the sample of size n given by

$$\mu_n(E) := \frac{1}{n} \sum_{i=1}^n \delta_{X_i}(E) = \frac{\#\{X_i \in E\}}{n} = \frac{1}{n} \sum_{i=1}^n \mathbf{1}\{X_i \in E\}, \quad E \in \mathcal{M}(\mathbb{R}), \quad (6)$$

where $\delta_a(\cdot) = \delta_0(\cdot - a)$ is the Dirac measure at a. The corresponding *empirical distribution function* F_n is the distribution function of μ_n

$$F_n(x) := \mu_n((-\infty, x]) = \frac{\#\{X_i \leq x\}}{n} = \frac{1}{n} \sum_{i=1}^n \mathbf{1}\{X_i \leq x\}, \quad x \in \mathbb{R}. \quad (7)$$

The strong law of large numbers (*Glivenko-Cantelli Theorem*) tells us that if F is the distribution function of the common distribution μ of X, we have for almost all

$\omega \in \Omega$

$$\sup_{x \in \mathbb{R}} |F_n(x, \omega) - F(x)| \to 0 \quad (n \to \infty) . \tag{8}$$

A proper scaling of this quantity leads to a non-trivial limit. Indeed if we fix $x \in \mathbb{R}$, the random variables $\mathbf{1}\{X_i \leqslant x\}$, $i \in \mathbb{N}$ are i.i.d. Bernoulli variables with $F(x) = \mathbb{P}(X_i \leqslant x)$. Now the central limit theorem (*de Moivre-Laplace Theorem*) states that

$$\sqrt{n}\,(F_n(x) - F(x)) = \frac{1}{\sqrt{n}} \sum_{i=1}^{n} \mathbf{1}\{X_i \leqslant x\} - F(x) \xrightarrow{d} \mathcal{N}\,(0, F(x)(1 - F(x))) . \tag{9}$$

This quantity can be viewed as a stochastic process indexed by $x \in \mathbb{R}$ which leads to the following definition.

Definition 4 Let $(X_i)_{i \in \mathbb{N}}$ be a sequence of i.i.d. random variables in $\mathbb{R}$ with common distribution function F and let F_n, $n \in \mathbb{N}$ be its empirical distribution function. We define the associated *empirical (error) process* by

$$G_n(x, \omega) := \sqrt{n}\,(F_n(x, \omega) - F(x)) , \quad 0 \leqslant x \leqslant 1 . \tag{10}$$

There is a huge literature on empirical processes, for an overview and more details we refer for instance to [2] or [22]. It is easily seen that G_n is a random element of the space $\mathbb{D}[0, 1]$, the space of càdlàg functions $\varphi : [0, 1] \to \mathbb{R}$, and the one dimensional (marginal) distributions of G_n are determined by (10). Moreover it is well known that the random variable $F(X)$ is uniformly distributed on the interval $[0, 1]$ whenever X has the continuous strictly increasing distribution function F, since in this case F is one to one and by an easy change of variables

$$\mathbb{P}(F(X) \in [a, b]) = \int_{F^{-1}(a)}^{F^{-1}(b)} f(x)dx = b - a \qquad \text{for all } 0 \leqslant a < b \leqslant 1.$$

However F is usually not known and to be estimated. In any case we have $F(X) \in [0, 1]$ whether F corresponds to the actual distribution of X or not. Hence distributions on the interval $[0, 1]$ are of particular interest.

Theorem 1 ([3]Theorem 14.3, p. 149) *Let $(X_i)_{i \in \mathbb{N}}$ be a sequence of i.i.d. random variables on $[0, 1]$ with common distribution function F and $(G_n)_{n \in \mathbb{N}}$ be the associated empirical processes. Then there exists a Gaussian random element G with values in $\mathbb{D}[0, 1]$ such that*

$$\sup_{x \in \mathbb{R}} |G_n(x, \omega) - G(x, \omega)| \xrightarrow{d} 0 \quad (n \to \infty).$$

Moreover G is the unique Gaussian process determined by $\mathbb{E}[G] = 0$ *and covariance*

$$\mathbb{E}[G_s \cdot G_t] = F(s)(1 - F(t)) \quad for\ 0 \leqslant s \leqslant t.$$

In particular if F is the uniform distribution on $[0, 1]$, *then* $G = B^0$ *is a* Brownian bridge, *that is a Brownian motion conditioned to end in* 0 *at time* $x = 1$.

3.2 Quantiles

In view of the Wasserstein distance it is also necessary to look at the *empirical quantile function* F_n^{-1}. For $n \in \mathbb{N}$ denote by $X_{i:n}$ the *i*-th *order statistic* of a sample of size *n*, i.e. the ordered sample

$$X_{i:1} \leqslant X_{i:2} \leqslant \cdots \leqslant X_{n:n}.$$

As F_n is discontinuous it is certainly not invertible yet the concept of quantiles for non invertible distributions allows to define for $0 < u \leqslant 1$

$$F_n^{-1}(u) = \inf\{x \in \mathbb{R} : F_n(x) \geqslant u\} = \inf\{x \in \mathbb{R} : \#\{X_i \leqslant x\} \geqslant nu\}$$
$$= \min\{X_{i:n} : \#\{X_{i:n} \leqslant x\} \geqslant nu\} = X_{\lceil nu \rceil:n} . \tag{11}$$

It is the left continuous inverse of the right continuous F_n. In analogy to (8) we have by the law of large numbers that

$$|F_n^{-1}(u, \omega) - F(u)^{-1}| \to 0 , \qquad \text{for all } u \in (0, 1) \text{ as } n \to \infty \text{ for almost all } \omega \in \Omega.$$

In general we cannot expect a uniform convergence in *u*, since for unbounded support of μ the values will be infinity. For a precise analogue to the Glivenko-Cantelli theorem one must therefore stay away from the endpoints of the open support of μ if they are positive or negative infinity. Following the analogy to the central limit theorem of formula (10) we introduce the empirical quantile process.

Definition 5 Let $(X_i)_{i \in \mathbb{N}}$ be a sequence of real valued i.i.d. random variables with common distribution function *F*. The *empirical quantile process* or simply *quantile process* is defined as

$$Q_n(u, \omega) := \sqrt{n}\left(F_n^{-1}(u, \omega) - F^{-1}(u)\right), \quad 0 \leqslant u \leqslant 1, n \in \mathbb{N} . \tag{12}$$

For uniform distributions we obtain an analogue of Theorem 1.

Theorem 2 *Let* $(X_i)_{i \in \mathbb{N}}$ *be a sequence of real valued i.i.d. random variables with common uniform distribution function* $F = U$. *Then there exists a Brownian bridge*

$(B_n^0)_{n\in\mathbb{N}}$ *such that*

$$\sup_{0\leq u\leq 1} |Q_n(u,\omega) - B_u^0(\omega)| \xrightarrow{d} 0. \tag{13}$$

As already mentioned for general distributions F uniform convergence of the quantile process is out of reach, instead we will consider cutoff L^2 distances later. In the following we link the empirical quantile process to Wasserstein distances.

3.3   The Empirical Wasserstein Distance

We can now calculate the Wasserstein distance of an empirical measure $\mu_n(\omega)$ to some given reference measure μ. We introduce the *Wasserstein statistic*

$$\mathrm{w_n}(\omega) := W_2^2(\mu_n(\omega), \mu). \tag{14}$$

We can calculate this distance by

$$\mathrm{w_n} = \int_0^1 |F_n^{-1}(u) - F^{-1}(u)|^2 du = \int_0^1 |X_{\lceil nu\rceil:n} - F^{-1}(u)|^2 du$$

$$= \sum_{i=1}^n \int_{\frac{i-1}{n}}^{\frac{i}{n}} (X_{(i-1):n} - F^{-1}(u))^2 du, \quad n \in \mathbb{N}. \tag{15}$$

The last expression turns out to be a quadratic polynomial in the order statistic, that is

$$\mathrm{w_n} = \sum_{i=1}^n a_i X_{i:n}^2 + b_i X_{i:n} + c, \tag{16}$$

where the coefficients are determined by the binomial formula and given by

$$a_i = \frac{1}{n}, \quad b_i = -2\int_{\frac{i-1}{n}}^{\frac{i}{n}} F^{-1}(u)du, \quad c = \int_0^1 \left(F^{-1}(u)\right)^2 du. \tag{17}$$

This formula can also be adapted for the trimmed version in the following sense. Consider the Wasserstein distance with respect to $|x|_1 = |x| \wedge 1$ on $\mathbb{R}$, which we shall denote by $\bar{W}_2$. As mentioned in Remark 2, $(F_n^{-1}(U), F^{-1}(U))$ is still a coupling, but in general not optimal with respect to $|\cdot|_1$. By definition $\bar{W}_2$ is the L^2-distance of the optimal coupling, and hence minimal. Therefore we obtain only the estimate

$$\bar{W}_2^2(\mu_n, \mu) \le \int_0^1 \left(|F_n^{-1}(u) - F^{-1}(u)|^2 \wedge 1 \right) du, \tag{18}$$

where the right-hand side can be calculated similarly to (16) as follows

$$w_n^* = \sum_{i=1}^n \beta_i^1 X_{i:n}^2 + \beta_i^2 X_{i:n} + \beta_i^3 + \beta^4, \tag{19}$$

where for the sequence $0 \le a_1^* \le b_1^* \le a_2^* \le \cdots \le a_n^* \le b_n^* \le 1$ given by

$$a_i^* = \left(\frac{i-1}{n} \vee |F(X_{i:n} - 1)| \right) \wedge \frac{i}{n}, \qquad b_i^* = \frac{i-1}{n} \vee \left(|F(X_{i:n} + 1)| \wedge \frac{i}{n} \right) \tag{20}$$

the coefficients are calculated by

$$\beta_i^1 = b_i^* - a_i^*, \quad \beta_i^2 = -2 \int_{a_i^*}^{b_i^*} F^{-1}(u) du, \quad \beta_i^3 = \int_{a_i^*}^{b_i^*} \left(F^{-1}(u) \right)^2 du, \quad \beta^4 = 1 - \sum_{i=1}^n \beta_i^1. \tag{21}$$

Note that the cutoff Wasserstein distance still can be written formally as a polynomial in $X_{i:n}$, since the monotonicity of F^{-1} yields that $(X_{i:n} - F^{-1}(u)) \wedge 1 = 0$ iff $u = F(X_{i:n} - 1)$. This allows to shift the condition with the minimum of the integrand of the coefficients β_i^j to the bounds of the integral a_i^* and b_i^*, which now also depend on $X_{i:n}$.

Remark 6 The concept of coupling distances was introduced in [9] for distributions in $\mathbb{R}^d$. However, in higher dimensions the concept of empirical quantile functions turns out to be much more involved.

3.4 Asymptotic Distribution and Rate of Convergence

For rigorous statistical applications it is necessary to determine the rate of convergence of the statistic of interest, in our case w_n^*. By definition w_n^* tends to zero. Quantifying the rate of convergence amounts to finding the correct renormalization to obtain a non-trivial (random) limit. For this purpose we need the notion of slow variation. A (measurable) function $\ell : (0, 1) \to (0, 1)$ is called *slowly varying* at zero (at one) if it satisfies for all $x \in (0, 1)$

$$\lim_{\substack{u \to 0 \\ (u \to 1)}} \frac{\ell(ux)}{\ell(u)} = 1.$$

The prototype of a slowly varying function at 0 is $\ell(u) = \ln(u)$. An easy calculation shows

$$\frac{\ell(ux)}{\ell(u)} = \frac{\ell(u)}{\ell(u)} + \frac{\ell(x)}{\ell(u)} \to 1 \text{ as } u \to 0+ .$$

We summarize the relevant properties. A classical reference is [4].

Proposition 1

1. Let ℓ be slowly varying at 0 (at 1). Then also ℓ^γ is slowly varying at 0 (at 1) for any $\gamma \in \mathbb{R}$.
2. For $\gamma > 0$ and ℓ slowly varying at 0, we have $u^\gamma \ell(u) \to 0$ and $u^{-\gamma} \ell(u) \to \infty$ as $u \to 0+$.

Let F be a distribution function on $\mathbb{R}$ with density f such that $f \circ F^{-1}$ is *regularly varying* at 0 and at 1, of index $\kappa_0, \kappa_1 > 1$ in the sense that

$$f(F^{-1}(u)) = u^{\kappa_0} \ell_0(u) \quad \text{on } (0, \tfrac{1}{2}], \quad \text{and} \quad f(F^{-1}(u)) = u^{\kappa_1} \ell_1(u) \quad \text{on } (\tfrac{1}{2}, 1)$$

Where $\ell_0(u)$ is slowly varying as $u \to 0$ and $\ell_1(u)$ as $u \to 1$. Let us further make the following technical assumptions

Assumptions 1

1. The interior of the support of f is an open interval.
2. f is continuously differentiable on its support apart from the boundary points.
3. The following uniform bound holds true

$$\sup_{0<u<1} u(1 - u) \frac{|f'(F^{-1}(u))|}{f^2(F^{-1}((u))} < \infty .$$

In the following we prove a polynomial rate of convergence of w_n^*.

Theorem 3 Assume that $|\kappa_0 - \kappa_1| < \tfrac{1}{2}$ and without loss of generality let $\kappa_0 \leqslant \kappa_1$. Choose now $\gamma_0 < \gamma < \gamma_1$ where

$$\gamma_0 := \max\left\{\frac{2\kappa_1 - 3}{2\kappa_1 - 2}, 0\right\} \quad \text{and} \quad \gamma_1 := \frac{2\kappa_0 - 2}{2\kappa_0 - 1} < 1. \tag{22}$$

Then there exists a sequence of non-negative random variables $(\Xi_{\gamma,n})_{n \in \mathbb{N}}$ (on the same probability space) such that for all $n \in \mathbb{N}$

$$n^{1-2(1-\gamma)(\kappa_0-1)} \, w_n{}^* \leq \Xi_{\gamma,n} \quad \mathbb{P} \text{ almost surely,}$$

and such that

$$\Xi_{\gamma,n} \xrightarrow{d} \int_0^1 (u^{\kappa_0-2} B_u)^2 du \quad as \quad n \to \infty ,$$

where $B = (B_u)_{u \in [0,1]}$ is a standard Brownian motion.

Proof Since $|\kappa_1 - \kappa_0| < \frac{1}{2}$ we are able to choose γ according to (22). Note that because $\frac{2\kappa_0-3}{2\kappa_0-1} \leq \gamma_0 < \gamma$ the renormalization factor $n^{1-2(1-\gamma)(\kappa_0-1)}$ diverges for n large. We then decompose our integral

$$\int_0^1 \left(|F_n^{-1}(u) - F^{-1}(u)|^2 \wedge 1 \right) du$$

$$= \left(\int_0^{n^{\gamma-1}} + \int_{n^{\gamma-1}}^{\frac{1}{2}} + \int_{\frac{1}{2}}^{1-n^{\gamma-1}} + \int_{1-n^{\gamma-1}}^1 \right) \left(|F_n^{-1}(u) - F^{-1}(u)|^2 \wedge 1 \right) du$$

$$\leq 2n^{\gamma-1} + \left(\int_{n^{\gamma-1}}^{\frac{1}{2}} + \int_{\frac{1}{2}}^{1-n^{\gamma-1}} \right) \left(|F_n^{-1}(u) - F^{-1}(u)|^2 \right) du . \qquad (23)$$

The integrals in (23) are treated with the help of Theorem 2.4 in [5]. Bearing in mind that $f(F^{-1}(u)) = u^{\kappa_0} \ell_0(u)$ for $u \in (0, \frac{1}{2}]$ it states for $k_n = n^\gamma$ and $Q_n(u) = \sqrt{n}(F_n^{-1}(u) - F^{-1}(u))$ (cf. Definition 5) that

$$\left(\frac{k_n}{n} \right)^{2(\kappa_0-1)} \ell_0^2 \left(\frac{k_n}{n} \right) \int_{\frac{k_n}{n}}^{\frac{1}{2}} Q_n(u)^2 du \xrightarrow{d} \int_0^1 u^{2\kappa_0-4} |B_u|^2 du \quad as \quad n \to \infty. \qquad (24)$$

On the other hand we assumed $\gamma < \gamma_1$ such that the first summand in

$$\Xi_{\gamma,n}^0 = n^{-2(1-\gamma)(\kappa_0-1)} \ell_0^2(n^{\gamma-1}) \left(nn^{\gamma-1} + \int_{n^{\gamma-1}}^{\frac{1}{2}} \left(\sqrt{n} |F_n^{-1}(u) - F^{-1}(u)| \right)^2 du \right) , \qquad (25)$$

tends to zero (after renormalization with the prefactor) and the whole expression has the above limit (24) . Analogous reasoning holds true for the remaining summands in (23) since $\gamma < \gamma_1 \leq \frac{2\kappa_1-2}{2\kappa_1-1}$. Hence

$$\Xi_{\gamma,n}^1 = n^{-2(1-\gamma)(\kappa_1-1)} \ell_1^2(n^{\gamma-1}) \left(nn^{\gamma-1} + \int_{\frac{1}{2}}^{1-n^{\gamma-1}} \left(\sqrt{n} |F_n^{-1}(u) - F^{-1}(u)| \right)^2 du \right) , \qquad (26)$$

has the same limit distribution as (24). So far we have obtained the convergence rate up to a regularly varying modulation. It is clear that since our estimate is not sharp, we can use the estimates for any larger γ to dominate the slowly varying factor. In fact for $\gamma < \gamma' < \gamma_1$ we have

$$n^{1-2(1-\gamma)(\kappa_0-1)} \, W_n{}^* \leqslant \left(\frac{n^{-2(1-\gamma)(\kappa_0-1)}}{n^{-2(1-\gamma')(\kappa_0-1)}\ell_0^2(n^{\gamma'-1})} \Xi_{\gamma',n}^0 + \frac{n^{-2(1-\gamma)(\kappa_0-1)}}{n^{-2(1-\gamma')(\kappa_1-1)}\ell_1^2(n^{\gamma'-1})} \Xi_{\gamma',n}^1 \right)$$

$$= c_0(n)\Xi_{\gamma',n}^0 + c_1(n)\Xi_{\gamma',n}^1$$

$$\leqslant \max\{c_0(n), 1\}\Xi_{\gamma',n}^0 + c_1(n)\Xi_{\gamma',n}^1 =: \Xi_{\gamma,n} .$$

The prefactors $c_0(n)$ and $c_1(n)$ vanish asymptotically for large n (see Proposition 1 (ii)) and $\Xi_{\gamma,n}$ has the same limit distribution as $\Xi_{\gamma',n}^0$.

Example 2 The assumptions on the underlying distribution are not restrictive.

1. Prominent examples of heavy tailed distributions are (symmetric) α-stable distributions. See [7]. They admit a density f and the distribution function F is regularly varying of order $-\alpha, \alpha \in (0, 2)$ at $\pm\infty$. L'Hospital's rule yields that f is also regularly varying of order $-(\alpha + 1)$. Obviously the quantile function F^{-1} is then regularly varying at 0 and 1 with index $-\frac{1}{\alpha}$ and

$$f(F^{-1}(u)) = u^{\frac{\alpha+1}{\alpha}} \ell(u)$$

for a function ℓ slowly varying at $\pm\infty$. It is an instructive exercise to verify Assumption 1 *(iii)*. In this case $\kappa = \kappa' = 1 + \frac{1}{\alpha}$ and we obtain a rate of convergence of our statistic of polynomial order

$$n^{1-2(1-\gamma)(\kappa-1)} \quad \text{with } 1 - \frac{\alpha}{2} < \gamma < \frac{1}{1+\frac{\alpha}{2}}.$$

Minimizing γ we can achieve any polynomial rate of convergence slower than $n^{\frac{1}{1+\frac{2}{\alpha}}}$. Since $\alpha \in (0, 2)$ this is slower than the order $\sqrt{n}$ of the central limit theorem.

2. The simplest examples of one-sided polynomial tails are Pareto distributions where

$$F(x) = 1 - \frac{c^\alpha}{x^\alpha}, \qquad x \geqslant c, \, \alpha > 1.$$

By analogous reasoning we can still achieve convergence rates arbitrarily close to $n^{\frac{1}{1+\frac{2}{\alpha}}}$, now including the values $\alpha > 2$. In this regime we are faster than the central limit theorem but always less than one.

Corollary 3.1 *In the situation of Theorem 3 our renormalized statistic* w_n^* *is asymptotically dominated by a limit with expectation* $\frac{1}{2\kappa-2}$ *and for* $\kappa > \frac{5}{4}$ *the variance is bounded by* $\frac{1}{4\kappa-5}$.

Proof Since $\mathbb{E}[B_u^2] = u$ we have by Fubini's theorem

$$\mathbb{E}[\int_0^1 u^{2\kappa-4}|B_u|^2 du] = \int_0^1 u^{2\kappa-3} du = \frac{1}{2\kappa-2},$$

and since $\frac{B_u}{\sqrt{u}} \sim N(0,1)$ we have

$$\mathbb{E}\left[\left(\int_0^1 (u^{\kappa-2}|B_u|)^2 du - \frac{1}{2\kappa-2}\right)^2\right] = \mathbb{E}\left[\left(\int_0^1 u^{2\kappa-3}\left(\frac{|B_u|^2}{u} - 1\right)du\right)^2\right]$$

$$\leq \int_0^1 \left[u^{2(2\kappa-3)}\mathbb{E}\left(\frac{|B_u|^2}{u} - 1\right)^2\right]du = 2\int_0^1 u^{2(2\kappa-3)} du = \frac{2}{2(2\kappa-3)+1}.$$

4 The Procedure in Detail

In this section we work through the program laid out in the previous sections. In order to keep calculations and the strategy easily tractable we will restrict ourselves to a simple example, which can be obviously adapted and extended. Our approach accounts for strong fluctuations in short time, which are hardly explained by a continuous evolution.

A reasonable modeling approach is hence to consider a process of the type

$$Y_t = G_t + L_t, \qquad t \in [0, T] \text{ for fixed } T > 0, \tag{27}$$

where $G = (G_t)_{t\in[0,T]}$ is a continuous process and $L = (L_t)_{t\in[0,T]}$ is a purely discontinuous Lévy process. We have seen in Sect. 2, that L is determined by a Lévy triplet of the form $(0, 0, \nu)$, where ν is a Lévy measure defined in (1). For instance, the solutions of stochastic differential equations

$$Y_t = x + \int_0^t f(Y_s)ds + L_t, \qquad t \in [0, T],$$

for globally Lipschitz continuous functions $f : \mathbb{R} \to \mathbb{R}$ fall into this class. The aim of this procedure is now to determine the nature of L and thus of its Lévy measure ν.

Given a data set $y = (y_i)_{i=0,...n}$ we interpret y as a realization of a process Y given in the class of models (27) observed at discrete times $t_1 < \cdots < t_i < \cdots < t_n$, that is

$$y_i = Y_{t_i}(\omega) \qquad \text{for some } \omega \in \Omega.$$

We make the following modeling assumptions:

Assumptions 2 *Fix a threshold $\rho > 0$.*

1. *The observation frequency is sufficiently high in comparison to the occurrence of large jumps given as increments beyond our threshold ρ. That means we assume that in each time interval $[t_{i-1}, t_i)$ at most one large jump occurs.*
2. *The behavior of small jumps is sufficiently benign in comparison to the large jump threshold ρ during our observation. That means in particular, we assume that over one time interval $[t_{i-1}, t_i)$ that small jump and continuous contributions cannot accumulate to this threshold.*

These assumptions can be made rigorous by further model assumptions on Y e.g. with the Lipschitz continuity of G. Under the assumptions it is justified to estimate

$$Y_{t_i} - Y_{t_{i-1}} \approx C_s^\rho - C_{s-0}^\rho \qquad \text{for exactly one } s \in [t_{i-1}, t_i),$$

for the compound Poisson process C^ρ given by (3) and hence can be considered as the realization of an i.i.d. sequence $X = (X_i)_{i=1,\dots,n}$. We denote by $x = (x_i)_{i=1,\dots n}$ the vector of large increments

$$x_i = (y_i - y_{i-1})\mathbf{1}\{|y_i - y_{i-1}| > \rho\}$$

this means

$$x_i = X_i(\omega) \qquad \text{for some } \omega \in \Omega.$$

Let μ_n be the empirical measure of the data X. The Glivenko-Cantelli theorem (7) tells us that for almost all $\omega \in \Omega$

$$\mu_n(\omega, \cdot) \to \mu \qquad n \to \infty, \text{ weakly}$$

for the common distribution μ of X. Since by construction $\mu = \nu_\rho$ we have

$$\mu_n(\omega) \to \mu \qquad n \to \infty, \text{ weakly, too.}$$

Since the Wasserstein distance encoded in the coupling distance metrizes the weak convergence we have for almost all $\omega \in \Omega$

$$T_{\lambda_\rho}(\mu_n, \nu_\rho) \to 0 \qquad \text{as } n \to \infty.$$

In particular we have an estimator for the tail of the Lévy measure ν_ρ. We are now in the position to estimate the distance between the Lévy measure of the compound Poisson approximation C^ρ of L (respectively Y) and the tail ν_ρ^* of a suspected reference Lévy measure ν^*. In particular

$$T_{\lambda_\rho}(\nu_\rho, \nu^*) \leqslant T_{\lambda_\rho}(\nu_\rho, \mu_n) + T_{\lambda_\rho}(\mu_n, \nu_\rho^*) \leqslant (\lambda_\rho \, \mathrm{w_n}^*(\mu_n, \nu^*))^{\frac{1}{2}} + T_{\lambda_\rho}(\nu_\rho, \mu_n), \tag{28}$$

where the second to last term tends to 0. The first term can be calculated explicitly due to Sect. 3. This is carried out in the case of polynomial tails in the next section.

5 Case Study

Many observed quantities in nature follow a heavy-tailed distribution, that can be interpreted as the superposition of (large) power law jumps. Small jumps are statistically hard to distinguish from continuous increments and will be neglected in this study. Therefore it is natural to consider jumps away from 0 with polynomial tails. In our simple model we assume

$$v(dy) = \frac{c_1 dy}{|y|^{1+\alpha_1}} \mathbf{1}\{y < -\rho_1\} + \frac{c_2 dy}{y^{1+\alpha_2}} \mathbf{1}\{y > \rho_2\}, \tag{29}$$

where $c_1, c_2 \geq 0$, $\alpha_1, \alpha_2 > 0$ and $\rho_1, \rho_2 > 0$ sufficiently small.

5.1 Concrete Formulas

To prevent numerical pathologies in the following calculation we exclude α_1 and α_2 from being 1 or 2. Due to Remark (1) we can restrict ourselves to one-sided Lévy measures and estimate the right and left tail independently, that is $c = c_1 > 0$, $c_2 = 0$, $\alpha = \alpha_1 > 0$ and $\rho = \rho_1 > 0$. Consider the Lévy measure v with Pareto density f given by

$$f(x) = \frac{c}{|x|^{\alpha+1}}, \qquad\qquad x \leq -\rho$$

$$F(x) = v((-\infty, x]) = \frac{c}{\alpha}|x|^{-\alpha}, \qquad\qquad x \leq -\rho$$

$$F^{-1}(u) = -(\frac{\alpha}{c}u)^{-1/\alpha}, \qquad\qquad u \in (0, 1).$$

Since the measure v is not a probability measure we will introduce the normalized measures v_λ of precisely mass $\lambda > 0$, that is supported on $(-\infty, -\rho]$, with

$$\rho = F^{-1}(\lambda) = \left(\frac{\alpha\lambda}{c}\right)^{-1/\alpha}. \tag{30}$$

We now can look at the normalization

$$\tilde{f}_\lambda(x) = \frac{c}{\lambda|x|^{\alpha+1}}, \quad x \leqslant -\rho$$

$$\tilde{F}_\lambda(x) = \frac{c}{\alpha\lambda}|x|^{-\alpha}, \qquad x \leqslant -\rho, \tag{31}$$

and its inverse

$$\tilde{F}_\lambda^{-1}(u) = -\left(\frac{\alpha\lambda}{c}u\right)^{-1/\alpha}, \qquad u \in (0,1).$$

In order to calculate the cutoff Wasserstein distance in formula (21) we evaluate the primitives

$$\int_z^1 \tilde{F}_\lambda^{-1}(u)\,du = -\frac{\alpha}{\alpha-1}\left(\frac{\alpha\lambda}{c}\right)^{-1/\alpha}\left(1-z^{1-1/\alpha}\right), \qquad z \in (0,1) \tag{32}$$

$$\int_z^1 \left(\tilde{F}_\lambda^{-1}(u)\right)^2 du = -\frac{\alpha}{\alpha-2}\left(\frac{\alpha\lambda}{c}\right)^{-2/\alpha}\left(1-z^{1-2/\alpha}\right), \qquad z \in (0,1). \tag{33}$$

Now all necessary functions to implement the empirical Wasserstein distance (19) are at our disposal. Recall the modeling Assumptions 2 and that the time series stems from a process of the form (27).

5.2 Simulations

In a first test case we simulate $n = 100.000$ data points from a perfectly symmetric version of the jump measure given in (29) with minimal jump sizes $\rho = 0.5$ and $\alpha^1 = 1.5$, $\alpha^2 = 1.8$, $\alpha^3 = 2.4$ and $\alpha^4 = 3.0$. We interpret those as the jumps of compound Poisson processes at rate $\lambda = 1$ and denote by μ_n^i the empirical measure of the respective simulation $i = 1,\ldots 4$. Furthermore we set $T = 1$ and apply a small linear drift $G(t) = 0.0125 \times t$ to each of them. The paths are shown in Fig. 1 (left). We calculate the empirical coupling distance between the simulated data and the jump measures according to the outlined procedure. The right display of Fig. 1 shows pronounced and small minima of the coupling distances as a function

$$\alpha \mapsto T_\lambda(v_\rho^\alpha, \mu_n^i) \in [0, \lambda^{\frac{1}{2}}]. \tag{34}$$

at the original values of $\alpha = \alpha_i$. Due to the cutoff of the Wasserstein distance at height 1 (see Definition 2) and the specific choice of the intensity $\lambda = 1$ the values of (34) are between 0 and 1. Due to Theorem 3, Corollary 3.1 and Example 2 we can estimate the expected value of the error $T_\lambda(v_\rho^{\alpha_i}, \mu_n^i)$. For large n this value (appropriately normalized) should be close to the expectation of the

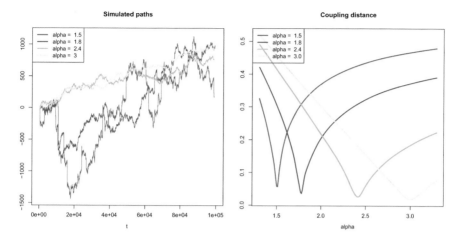

Fig. 1 *Left*: Simulated paths for fixed values of α. *Right*: Empirical coupling distances between the data of the simulated paths and the jump measures (31), as a function of varying α

limiting distribution. For the optimal rate of convergence, $\gamma = 1 - \frac{\alpha}{2}$, this expression is given by

$$E^{i,n} := \left(\frac{\mathbb{E}[\int_0^1 (u^{1+\frac{1}{\alpha}} B_u)^2 du]}{\lambda \, n^{1+\frac{2}{\alpha}}} \right)^{\frac{1}{2}} = \sqrt{\frac{\alpha}{2\lambda}} \, n^{-\frac{1}{2(1+\frac{2}{\alpha})}}. \tag{35}$$

The simulation results are compared to the expected value in the following table.

α_i	1.5	1.8	2.4	3.0
$T_\lambda(\nu_\rho^{\alpha_i}, \mu_n^i)$	0.0575	0.0642	0.0273	0.0183 $\ll 1$
$E^{i,n}$	0.0735	0.0621	0.0474	0.0387

Apparently the empirical results are better than our prediction $E^{i,n}$, that is the empirical errors from the data are somewhat smaller than the estimated error. Recall that the optimal rate in Example 2 is not achieved, hence we renormalize by a too small value in formula (35) and consequently slightly overestimate the value of our statistic $T_\lambda(\nu_\rho^{\alpha_i}, \mu_n^i)$.

5.3   Paleoclimatic Time Series

The concentration of calcium ions in ice core data from the Greenland shelf provide a climate proxy for the yearly average temperature distribution during the last glacial period, see [17]. The record shows large fluctuations between (cooler) stadials and (warmer) interstadials, see Fig. 2 (left up). In [6] Ditlevsen concentrated on the actual (42) transitions between the different regimes which he interprets as the effect of single large jumps of an underlying noise signal. His analysis indicates a tail index of $\alpha = 1.75$. Moreover he proposed an α-stable Lévy component in the noise signal.

A series of works [10, 13] and [12] continued the investigation on the mathematical side towards α-stable perturbations and developed an estimation procedure based on the selfsimilarity and characteristic path variations. In the realm of stable diffusions their method proposes the value of $\alpha = 0.75$ as most likely, and $\alpha = 1.75$ as reasonable alternative. In [8] Wasserstein distances were also applied to measure the distance between α-stable distributions and empirical measures, however due to the lack of moments the analysis is restricted to W_p for $p = p(\alpha) < 1$.

These elaborate techniques are hardly applicable beyond this framework. The procedure proposed here does not rely on any features of stable distributions and only requires a monotonic tail behavior of the jump distribution. There is a variety of estimators for the tail index in the literature, our method does not intend to contribute to this list. Instead our method quantifies the fit of a *proposed* tail behavior to a *given* data set. The proposed tails could be derived by standard methods. Certainly coupling distances can also serve in a minimal distance estimation procedure. In order to make such a procedure statistically rigorous it would be necessary to

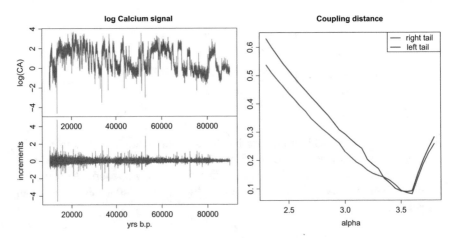

Fig. 2 *Left (above)*: Logarithmic calcium signal. *Left (below)*: Increments of the logarithmic calcium signal. *Right*: Plot of the coupling distance for the left tail (*blue*) and the right tail (*red*) of the jump measure

further quantify the separation power by finding uniform lower bound jointly in the parameters (α, λ, ρ).

Our model only describes large jumps, small fluctuations can certainly not be described by the (polynomial) tail behavior. In other words, the modeling Assumptions 2 are valid only for increments beyond a certain threshold $\rho > 0$. Small fluctuations are interpreted as contributions of a continuous part G and small jumps as in (27). Also note that for $\alpha > 2$ in the limit of ρ to 0 the measure ν from (29) is not square integrable in the neighborhood of 0. Hence it is not a Lévy measure (cf. formula (1) in Sect. 2).

Our procedure shows marked minima of the coupling distance for the right and the left side of the one-sided polynomial tails given in (29) as a function in α for $\alpha_1 = 3.55$ (left) and $\alpha_2 = 3.6$ (right). The Wasserstein distances, which correspond to the coupling distance to the interval $[0, 1]$ gives the small values $0.089 \ll 1$ (left) and $0.081 \ll 1$ (right). The cutoffs have been chosen $\rho_1 = 0.34$ allowing for $n_1 = 530$ sample points and $\rho_2 = 0.36$ with $n_2 = 894$ sample points. The respective rates are $\lambda_{\rho_1} = 11$ and $\lambda_{\rho_2} = 8$. The prediction procedure developed above yields $E^{1,n_1} = 0.054$ and $E^{2,n_2} = 0.053$. This under estimation could be explained beyond climatological reasoning by comparably small sample sizes, where the asymptotic regime is not yet fully unfolded. Yet the order of magnitude is caught.

We point out that there is a significant degree of freedom in the 3 parameter fit. For instance the cutoff parameter may vary between the floating boundaries of the modeling assumption of a meaningful tail contribution (not corrupted by small scale fluctuations) and a statistically relevant number of data points.

Comparable fits for $\alpha = 1.75$ with a significant number of data points $n > 100$ do not allow for reasonably small Wasserstein distances (all values > 0.5). Consequently we could not confirm the suggestion of a tail index $\alpha = 1.75$ in the literature. The proposed procedure indicates a Lévy jump component with a tail index clearly above 2, that cannot belong to the family of stable diffusions.

Acknowledgements The authors would like to thank Sylvie Roelly and University Potsdam for constant hospitality and support. J. Gairing and M. Högele would like to thank the IRTG 1740 Berlin-São Paulo: "Dynamical Phenomena in Complex Networks" and the Berlin Mathematical School (BMS) for infrastructure support. The DFG Grant Schi 419/8-1 and the joint DFFD-RFFI project No. 09-01-14 are gratefully acknowledged by A. Kulik.

References

1. Applebaum, D.: Lévy Processes and Stochastic Calculus. Cambridge University Press, Cambridge (2009)
2. del Barrio, E., Deheuvels, P., van de Geer, S.: Lectures on Empirical Processes. EMS Series of Lectures in Mathematics. EMS, Zürich (2007)
3. Billingsley, P.: Convergence of Probability Measures. Wiley Series in Probability and Statistics. Wiley, New York (1999)
4. Bingham, N.H., Goldie, C.M., Teugels, J.L.: Regular Variation. Cambridge University Press, Cambridge (1987)

5. Csörgö, M., Horváth, C.: On the distributions of Lp norms and weighted quantile processes. Annales de l'I.H.P. Section B **26**(1), 65–85 (1990)
6. Ditlevsen, P.D.: Observation of a stable noise induced millennial climate changes from an ice-core record. Geophys. Res. Lett. **26**(10), 1441–1444 (1999)
7. Feller, W.: An Introduction to Probability Theory and its Applications. Geophysical Research Letters, vol. II. John Wiley & Sons, New York (1971)
8. Gairing, J.: Speed of convergence of discrete power variations of jump diffusions. Diplom thesis, Humboldt-Universität zu Berlin (2011)
9. Gairing, J., Högele, M., Kosenkova, T., Kulik, A.: Coupling distances between Lévy measures and applications to noise sensitivity of SDE. Stoch. Dyn. **15**(2), 1550009-1–1550009-25 (2014). doi:10.1142/S0219493715500094, online available
10. Gairing, J., Imkeller, P.: Stable CLTs and rates for power variation of α-stable Lévy processes. Methodol. Comput. Appl. Probab. **17**, 1–18 (2013)
11. Debussche, A., Högele, M., Imkeller, P.: The Dynamics of Non-linear Reaction-Diffusion Equations with Small Lévy Noise. Springer Lecture Notes in Mathematics, vol. 2085. Springer, Cham (2013)
12. Hein, C., Imkeller, P., Pavlyukevich, I.: Limit theorems for p-variations of solutions of SDEs driven by additive stable Lévy noise and model selection for paleo-climatic data. Interdisciplinary Math. Sci. **8**, 137–150 (2009)
13. Hein, C., Imkeller, P., Pavlyukevich, I.: Simple SDE dynamical models interpreting climate data and their meta-stability. Oberwolfach Reports (2008)
14. Högele, M., Pavlyukevich, I.: The exit problem from the neighborhood of a global attractor for heavy-tailed Lévy diffusions. Stoch. Anal. Appl. **32**(1), 163–190 (2013)
15. Imkeller, P., Pavlyukevich, I.: First exit times of SDEs driven by stable Lévy processes. Stoch. Process. Appl. **116**(4), 611–642 (2006)
16. Imkeller, P., Pavlyukevich, I., Wetzel, T.: First exit times for Lévy-driven diffusions with exponentially light jumps. Ann. Probab. **37**(2), 530–564 (2009)
17. Members NGRIP: NGRRIP data. Nature **431**, 147–151 (2004)
18. Kosenkova, T., Kulik, A.: Explicit bounds for the convergence rate of a sequence of Markov chains to a Lévy-type process. (in preparation)
19. Pavlyukevich, I.: First exit times of solutions of stochastic differential equations with heavy tails. Stoch. Dyn. **11**, Exp.No.2/3:1–25 (2011)
20. Rachev, S.T., Rüschendorf, L.: Mass Transportation Problems. Vol. I: Theory, Vol. II.: Applications. Probability and its Applications. Springer, New York (1998)
21. Samoradnitsky, G., Taqqu, M.: Stable non-Gaussian random processes. Chapman& Hall, Boca-Raton/London/New York/Washington, DC (1994)
22. van der Vaard, A.W., Wellner, J.A.: Weak Convergence and Empirical Processes. Springer Series in Statistics. Springer, New York (1996)

Part IV
Data Assimilation

Data Assimilation for Geophysical Fluids: The Diffusive Back and Forth Nudging

Didier Auroux, Jacques Blum, and Giovanni Ruggiero

Abstract Data assimilation is the domain at the interface between observations and models, which makes it possible to identify the global structure of a geophysical system from a set of discrete space-time data. After recalling state-of-the-art data assimilation methods, the variational 4DVAR algorithm and sequential Kalman filters, but also the Back and Forth Nudging (BFN) algorithm, we present the Diffusive Back and Forth Nudging (DBFN) algorithm, which is a natural extension of the BFN to some particular diffusive models. We numerically test the DBFN and compare it with other methods on both a 2D shallow water model, and a 3D full primitive ocean model. These numerical results show that the DBFN converges very quickly. Its main advantage is to filter the observation noise, unlike the BFN algorithm, particularly when the model diffusion is high.

Keywords Data assimilation • Geophysics • Nudging • Oceanography • Shallow water • Primitive equations • Back and forth nudging

1 Introduction

It is well established that the quality of weather and ocean circulation forecasts is highly dependent on the quality of the initial conditions. Geophysical fluids (air, atmospheric, oceanic, surface or underground water) are governed by the general equations of fluid dynamics. Geophysical processes are hence non linear because of their fluid component. Such non linearities impose a huge sensitivity to the initial conditions, and then an ultimate limit to deterministic prediction (estimated to be about 2 weeks for weather prediction for example). This limit is still far from being reached, and substantial gain can still be obtained in the quality of forecasts.

D. Auroux (✉) • J. Blum
Laboratoire J.A. Dieudonné, Université de Nice Sophia Antipolis, Parc Valrose, 06108 Nice cedex 2, France
e-mail: auroux@unice.fr; jblum@unice.fr

G. Ruggiero
Mercator Océan, 10 Rue Hermès, 31520 Ramonville-Saint-Agne, France
e-mail: giovanni.ruggiero@mercator-ocean.fr

© Springer International Publishing Switzerland 2016
F. Ancona et al. (eds.), *Mathematical Paradigms of Climate Science*, Springer
INdAM Series 15, DOI 10.1007/978-3-319-39092-5_8

This can be obtained through improvement of the observing system itself, but also through improvement of the geophysical models used to modelize the geophysical processes. For example, a major problem comes from the fact that sub-scaled processes could be associated with extremely large fluxes of energy. Seeking a numerical solution to the equations requires discretizing the equations, and therefore cutting off in the scales. It will be crucial to represent the fluxes of energy associated to sub-grid processes by some additional terms in the equations [36, 50].

Over the past 20 years, observations of ocean and atmosphere circulation have become much more readily available [13], as a result of new satellite techniques and international field programs (MERCATOR, CLIPPER, GODAE, ARGO, ...). In the case of the ocean modelling, the use of altimeter measurements has provided extremely valuable information about the sea-surface height, and then has allowed the oceanographic community to study more precisely both the general circulation of the ocean and the local dynamics of some particular regions (the Gulf Stream area, for example, but also the Kuroshio extension, the Antarctic circumpolar current and the tropical oceans). Geostationnary satellites also provide information on the wind by estimating the shifting of clouds considered as lagrangian tracers. Polar orbiting satellites are used for the estimation of the atmospheric vertical temperature profiles. Generally, radiances are measured and then temperatures are estimated as the solution of an inverse problem.

Meteorologic and oceanographic data are currently extremely heterogeneous, both in nature, density and quality, but their number is still smaller than the degree of freedom of the models. The growth of the available computing ressources indeed allows refinements of the grid size of general circulation models.

Environmental scientists are increasingly turning to inverse methods for combining in an optimal manner all the sources of information coming from theory, numerical models and data. Data assimilation (DA) is precisely the domain at the interface between observations and models which makes it possible to identify the global structure of a system from a set of discrete space-time data. DA covers all the mathematical and numerical techniques in which the observed information is accumulated into the model state by taking advantage of consistency constraints with laws of time evolution and physical properties, and which allow us to blend as optimally as possible all the sources of information coming from theory, models and other types of data.

There are two main categories of data assimilation techniques [58], variational methods based on the optimal control theory [42] and statistical methods based on the theory of optimal statistical estimation (see, for example, [12, 13, 39] for an overview of inverse methods, both for oceanography and meteorology). The first class of methods (3D-VAR, 4D-VAR, 4D-PSAS, ...) was first introduced in meteorology [19, 41, 59] and more recently for oceanic data [45, 48, 49, 54, 55, 60]. The statistical (or sequential) methods (optimal interpolation, Kalman filter, SEEK filter, ...) were introduced in oceanography roughly 15 years ago [30, 32]. The Kalman filter was extended to nonlinear cases [29, 38] but it has been mostly applied in oceanography to quasi-linear situations, in particular tropical oceans [16, 26, 27, 34, 62].

In practice, all data assimilation techniques encounter major difficulties due to computational reasons. The full Kalman filter would, in principle, require the manipulation of matrices with a dimension of typically 10^7 or 10^8 in an oceanic problem. The optimal control adjoint method often requires several hundred iterations of the minimization process to converge, thus implying an equivalent number of model runs. In this context, it is important to find new data assimilation algorithms allowing in particular a reduction of the problem dimension.

In this paper, we focus our interest on various data assimilation algorithms in order to identify the initial condition of a geophysical system and reconstruct its evolution in time and space.

We first study in Sect. 2 the four dimensional variational adjoint method (named 4D-VAR), using a strong constraint hypothesis (the ocean circulation model is assumed to be exact). The use of a cost function, measuring the mean-square difference between the observations and the corresponding model variables, allows us to carry out the assimilation process by an identification of the initial state of the ocean which minimizes the cost function.

Sequential methods are mostly based on the Kalman filtering theory, which consists in a forecast step and an analysis (or correction) step. In Sect. 3, we present the extended Kalman filter (EKF), for nonlinear models. A main drawback of the (extended) Kalman filter is the computational cost of propagating in time the error covariance matrices. We present then the ensemble Kalman filter (EnKF), for which an ensemble of states (members) is used to compute actual covariance matrices at a lower computational cost.

We recall in Sect. 4 the Back and Forth Nudging (BFN) algorithm, which is the prototype of a new class of data assimilation methods, although the standard nudging algorithm is known for a couple of decades. It consists in adding a feedback term in the model equations, measuring the difference between the observations and the corresponding space states. The idea is to apply the standard nudging algorithm to the backward (in time) nonlinear model in order to stabilize it. The BFN algorithm is an iterative sequence of forward and backward resolutions, all of them being performed with an additional nudging feedback term in the model equations. We also present the Diffusive Back and Forth Nudging (DBFN) algorithm, which is a natural extension of the BFN to some particular diffusive models. This section ends with theoretical considerations on both BFN and DBFN algorithms.

In Sect. 5, we study the behaviour of the DBFN algorithm on a 2D shallow water model. We report the results of numerical experiments, showing the convergence of the DBFN algorithm, a comparison between BFN, DBFN and 4DVAR algorithms, and sensitivity studies with respect to observation noise, and to the model diffusion coefficient. Note that we compare the BFN and DBFN algorithms with the 4DVAR algorithm, as they all consist in iterative resolutions of forward and backward (or adjoint) models, and they also all update the initial condition at each iteration. Kalman filter methods are sequential and do not update the initial condition. This is why we decided not to compare numerically BFN and DBFN algorithms with a Kalman filter approach.

In Sect. 6, we present numerical results on a full primitive ocean model (NEMO). After showing the convergence of the DBFN algorithm, we compare the BFN and DBFN algorithms.

Finally, some concluding remarks and perspectives are shown in Sect. 7.

2   Variational Method: 4D-VAR

Variational methods consider the equations governing the geophysical flow as constraints, and the problem is closed by using a variational principle, e.g. the minimization of the discrepancy between the model and the observations.

2.1   Model and Observations

Every DA method needs both a model describing the evolution of the fluid, basically a system of non linear partial differential equations (PDE), and a set of discrete observations. Firstly, we assume that the model can be written, after discretization in space of the set of PDE:

$$\begin{cases} \dfrac{dX}{dt} = F(X, U), \quad 0 < t < T, \\ X(0) = V, \end{cases} \tag{1}$$

where X is the state variable which describes the evolution of the system at each grid point. X depends on time, and is for operational models of large dimension (10^7–10^8). F is a non linear differential operator, describing the dynamics of the system. U corresponds to some internal variables of the model (parameters or boundary conditions) and may be time dependent. Finally, V is the initial condition of the system state, which is unknown. In order to use optimal control techniques, we have to define a control variable that should be identified. Most of the time, the control is (U, V), the initial condition and the model parameters.

Secondly, we suppose that we have an observation vector X_{obs} which gathers all the data we want to assimilate. These observations are discrete in time and space, distributed all over the assimilation period $[0, T]$, and are not in the same space as the state variable, from a geographical or a physical point of view. Therefore, we will need an observation operator C mapping the space of state into the space of observations. This operator can be non linear in some cases.

2.2 Cost Function

It is now possible to define a cost function $\mathscr{J}$ measuring the discrepancy of the solution of the model associated to the control vector (U, V) and the observations X_{obs}:

$$\mathscr{J}(U, V) = \frac{1}{2} \int_0^T \langle R^{-1}(CX - X_{obs}), CX - X_{obs} \rangle dt$$
$$+ \frac{1}{2} \langle P_0^{-1} V, V \rangle + \frac{1}{2} \int_0^T \langle Q^{-1} U, U \rangle dt \qquad (2)$$

where X is the solution of (1). R, P_0 and Q are covariance matrices, allowing us to introduce some a priori information about the statistics of the fields X_{obs}, V and U respectively. $\langle ., . \rangle$ is most of the time the canonical real scalar product.

The first part of the cost function quantifies the difference between the observations and the state function, and the two others act like a regularization term in the sense of Tykhonov. It is sometimes replaced by the so-called background term, which is the quadratic (with respect to the covariance matrix norm) difference between the initial optimal variable and the last prediction [19].

The inverse problem which consists in the minimization of the cost function $\mathscr{J}$ is then generally well-posed. The variational formulation of our DA problem can then be written as:

$$\begin{cases} \text{Find}(U^*, V^*) \text{ such that} \\ \mathscr{J}(U^*, V^*) = \inf_{(U,V)} \mathscr{J}(U, V). \end{cases} \qquad (3)$$

2.3 Gradient Step

In order to minimize the cost function, we need its gradient $\nabla \mathscr{J}$. Because of the large dimension of the model state vector (usually more than 10^7), it is not possible to compute directly the gradient by using finite difference methods. The gradient vector of the functional is then obtained by the adjoint method [18, 19]. Let $\hat{X}$ be the derivative of X with respect to (U, V) in the direction (u, v). Then $\hat{X}$ is solution of the following set of discretized partial differential equations, known as the linear tangent model:

$$\begin{cases} \dfrac{d\hat{X}}{dt} = \dfrac{\partial F}{\partial X} \hat{X} + \dfrac{\partial F}{\partial U} u, \\ \hat{X}(0) = v, \end{cases} \qquad (4)$$

where $\dfrac{\partial F}{\partial X}$ and $\dfrac{\partial F}{\partial U}$ represent the jacobian of the model with respect to the state variable and the model parameters respectively.

If we assume that the operator C is linear (otherwise, we have to linearize it), the derivative of $\hat{\mathscr{J}}$ with respect to (U, V) in the direction (u, v) is then

$$\langle \hat{\mathscr{J}}(U, V), (u, v)\rangle = \int_0^T \langle R^{-1}(CX - X_{obs}), C\hat{X}\rangle dt$$

$$+ \langle P_0^{-1}V, v\rangle + \int_0^T \langle Q^{-1}U, u\rangle dt.$$

We can introduce the so-called adjoint state P (which lives in the same space as X), solution of the adjoint model [19]:

$$\begin{cases} -\dfrac{dP}{dt} = \left(\dfrac{\partial F}{\partial X}\right)^T P - C^T R^{-1}(CX - X_{obs}), \\ P(T) = 0. \end{cases} \tag{5}$$

We have then:

$$\langle \hat{\mathscr{J}}(U, V), (u, v)\rangle = \int_0^T \langle \dfrac{dP}{dt} + \left(\dfrac{\partial F}{\partial X}\right)^T P, \hat{X}\rangle dt$$

$$+ \langle P_0^{-1}V, v\rangle + \int_0^T \langle Q^{-1}U, u\rangle dt$$

and an integration by part shows that, using (4):

$$\langle \hat{\mathscr{J}}(U, V), (u, v)\rangle = \int_0^T \langle -P, \dfrac{\partial F}{\partial U}u\rangle dt - \langle P(0), v\rangle$$

$$+ \langle P_0^{-1}V, v\rangle + \int_0^T \langle Q^{-1}U, u\rangle dt.$$

Finally, the gradient of $\mathscr{J}$ is given by:

$$\nabla \mathscr{J}(U, V) = \begin{pmatrix} -\left(\dfrac{\partial F}{\partial U}\right)^T P + Q^{-1}U \\ -P(0) + P_0^{-1}V \end{pmatrix}. \tag{6}$$

Therefore, the gradient is obtained by a backward integration of the adjoint model (5), which has the same computational cost as one evaluation of $\mathscr{J}$.

2.4   Optimality System

The minimization problem (3) is then equivalent to the following optimality system:

$$\begin{cases} \dfrac{dX}{dt} = F(X, U^*), \\ X(0) = V^*, \end{cases}$$

$$\begin{cases} -\dfrac{dP}{dt} = \left(\dfrac{\partial F}{\partial X}\right)^T P - C^T R^{-1}(CX - X_{obs}), \\ P(T) = 0, \end{cases} \tag{7}$$

$$\left(\dfrac{\partial F}{\partial U}\right)^T P = Q^{-1} U^*,$$

$$P(0) = P_0^{-1} V^*.$$

2.5   4D-Var Algorithm Computation

The determination of (U^*, V^*), solution of (3) and (7), is carried out by running a descent-type optimization method. We may use as a first guess (U_0, V_0) the result of the minimization process at the last prediction. Then, given the first guess, we use an iterative algorithm [33]:

$$(U_n, V_n) = (U_{n-1}, V_{n-1}) - \rho_n D_n$$

where D_n is a descent direction, and ρ_n is the step size.

The knowledge of (U_{n-1}, V_{n-1}) allows us to compute the corresponding solution X_{n-1} of the direct model (1), and consequently to evaluate the cost function $\mathcal{J}(U_{n-1}, V_{n-1})$. Then we solve the adjoint model (5) and compute the adjoint solution P_{n-1}, and using (6), the gradient of the cost function $\nabla \mathcal{J}(U_{n-1}, V_{n-1})$. The computation of the descent direction D_n is usually performed using conjugate gradient or Newton type methods. Finally, the step size ρ_n is chosen to be the step size which minimizes

$$\mathcal{J}((U_{n-1}, V_{n-1}) - \rho D_n)$$

with respect to ρ. This is a one-dimensional minimization, but in case the problem is non linear, we can get a high computational cost because it will require several evaluations of $\mathcal{J}$, and hence several integrations of the model (1) [15, 33, 43, 61].

2.6 Computational Issues

One of the most difficult steps in the 4D-Var algorithm is the implementation of the adjoint model. Numerically, the goal is to solve the discrete optimality system, which gives the solution of the discrete direct problem, and the discrete gradient is given by the discrete adjoint model, which has to be derived from the discrete direct model, and not from the continuous adjoint model. A bad solution would be to derive the adjoint model from the continuous direct model, and then to discretize it. The good solution is to first derive the tangent linear model from the direct model. This can be done by differentiating the direct code line by line. And then one has to transpose the linear tangent model in order to get the adjoint of the discrete direct model. To carry out the transposition, one should start from the last statement of the linear tangent code and transpose each statement. The derivation of the adjoint model can be long. Sometimes, it is possible to use some automatic differentiation codes (the direct differentiation gives the tangent linear model, and the inverse differentiation provides the adjoint model) [47, 52].

Another issue is the relative ill-posedness of the problem when the model is non linear. The cost function $\mathcal{J}$ is hence non convex, and may have plenty of local minima. The optimization algorithm may then converge toward a local minimum and not the global minimum. For this reason, the choice of the initial guess is extremely important, because if it is located in the vicinity of the global minimum, one can expect a convergence toward the global minimum. Another solution is to increase the weight of the two last terms of $\mathcal{J}$ in (2), which correspond to two regularization terms with respect to the two control variables. This has to be done carefully because it can provide a physically incorrect solution: if P_0 and Q are too small, the regularization of $\mathcal{J}$ is indeed a penalization. But usually, these regularization terms are used to force the model to verify some additional physical constraints or/and to take into account some statistical information on model/observation/background errors.

3 Sequential Methods: Kalman Filter

In this section, we will study data assimilation methods based on the statistical estimation theory, in which the Kalman filtering theory is the primary framework. But the application of this theory encounters enormous difficulties due to the huge dimension of the state vector of the considered system. A further major difficulty is caused by its non linear nature. To deal with this, one usually linearizes the ordinary Kalman filter (KF) leading to the so-called extended Kalman filter (EKF) [22, 28, 31, 62]. We will also present the ensemble Kalman filter (EnKF), which allows one to get rid of too expensive computations of covariance matrices.

3.1   The Extended Kalman Filter

Consider a physical system described by

$$X(t_i) = \mathcal{M}(t_{i-1}, t_i)X(t_{i-1}) + U_i \tag{8}$$

where $\mathcal{M}(t_{i-1}, t_i)$ is an operator describing the system transition from time t_{i-1} to t_i, usually obtained from the integration of a partial differential system, and U_i is an unknown term of the model (it can be a noise term, used to modelize the unknown parameters of the model [17]). We suppose that at each time t_i, we have an observation vector $X_{obs}(t_i)$. Let us denote by ε_i the observation error, i.e. the difference between the observation vector and the corresponding state vector:

$$\varepsilon_i = X_{obs}(t_i) - C_i X(t_i), \tag{9}$$

where C_i is the observation operator at time t_i, mapping the state space into the space of observations. Q_i and R_i will be the covariance matrices of the model error (U_i) and the observation error (ε_i) respectively.

The extended Kalman filter operates sequentially: from an analysis state vector $X_a(t_{i-1})$ and its error covariance matrix $P^a(t_{i-1})$, it constructs the next analysis state vector $X_a(t_i)$ and $P^a(t_i)$ in two steps, a forecasting step and a correction step.

The first step is used to forecast the state at time t_i:

$$X^f(t_i) = M(t_{i-1}, t_i)X^a(t_{i-1}), \tag{10}$$

where $M(t_{i-1}, t_i)$ is the linearized model around $X^a(t_{i-1})$. The forecast error covariance matrix is then approximately

$$P^f(t_i) = M(t_{i-1}, t_i)P^a(t_i)M(t_{i-1}, t_i)^T + Q_i. \tag{11}$$

The second step is an analysis step, the newly available observation $X_{obs}(t_i)$ is used to correct the forecast state vector $X^f(t_i)$ in order to define a new analysis vector:

$$X^a(t_i) = X^f(t_i) + K_i(X_{obs}(t_i) - C_i X^f(t_i)), \tag{12}$$

where K_i is a gain matrix, called the Kalman matrix. The optimal gain is given by

$$K_i = P^f(t_i)C_i^T \left(C_i P^f(t_i)C_i^T + R_i\right)^{-1}. \tag{13}$$

The corresponding analysis error covariance matrix is given by

$$P^a(t_i) = P^f(t_i) - P^f(t_i)C_i^T \left(C_i P^f(t_i)C_i^T + R_i\right)^{-1} C_i P^f(t_i). \tag{14}$$

One main issue of the EKF is that the covariance matrices R_i, Q_i and P_0^a have to be known. Some statistical information can be obtained for observation error from the knowledge of the instrumental error variances in situations such as altimetric observations from satellites over the ocean, for which the error estimates have become fairly solidly established. But it is not clear how the correlations of these errors can be obtained. The covariances matrices Q_i and P_0^a are much more difficult to obtain, because very little is known concerning the true initial state of the system. These matrices are of very large dimension, and usually have a quite large number of independent elements. Is it really useful to estimate such a huge number of parameters? The theory for such equations (Eqs. (11) and (14)) state that for linear autonomous systems, even if P_0^a is poorly specified, one may hopefully still have a good approximation to P_i^a in the long term. The Kalman filter is optimal only if the covariance matrices R_i and Q_i are correctly specified. Thus, in practice, the Kalman filter is suboptimal.

3.2 Ensemble Kalman Filter

A main issue of the (extended) Kalman filter is the computational cost of propagating the covariance matrices in time. The dimension of P^f and P^a matrices is usually too large in geophysical problems, and there are several ways to avoid this point. One interesting approach is the ensemble Kalman filter (EnKF). The EnKF can be seen as a Monte Carlo approximation of the KF, avoiding evolving the full covariance matrix of the probability density function of the state vector [13, 23–25, 37].

As error statistics of background errors are not very well known, the idea is to generate a set of perturbed background states, with small perturbations around the background state with the same probability distribution. For $1 \leq j \leq M$, M being the size of the ensemble, we define the background ensemble members:

$$X_j(t_0) = X^b(t_0) + \varepsilon_j, \tag{15}$$

where X^b is the background state, and ε_j is the statistical perturbation, with a probability distribution consistent with the background error covariance matrix.

Then, we obtain an ensemble of forecast states with

$$X_j^f(t_i) = \mathcal{M}(t_{i-1}, t_i) X_j^a(t_{i-1}), \tag{16}$$

where we use the nonlinear model $\mathcal{M}(t_{i-1}, t_i)$. The correlation between these states gives some information about the forecast error statistics. The forecast error covariance matrix is then the actual covariance of the ensemble of states.

Then, the analysis state step is similar to the standard Kalman filter, with the computation of the Kalman gain matrix, and the correction is applied to each member:

$$X_j^a(t_i) = X_j^f(t_i) + K_i(X_{obs}(t_i) - C_i X_j^f(t_i)), \tag{17}$$

where K_i is the Kalman gain matrix computed with the actual covariance matrices of the ensemble. Then, the analysis error covariance matrix is computed from the ensemble of analysis states.

The EnKF is then *simply* a Kalman filter, applied to a discrete set (ensemble) of states (members). The covariance matrices are computed from this set, without using the linear or adjoint model. There are two main advantages: first, the computational cost is much lower, as there are no costly computations for the covariance matrices; and second, the covariance matrices represent the actual covariance statistics of the members in the ensemble.

4   Back and Forth Nudging Schemes: BFN and Diffusive BFN (DBFN)

The main issues of data assimilation for geophysical systems are the huge dimension of the control vectors (and hence of the covariance matrices) and the non linearities (most of the time, one has to linearize the model and/or some operators). The computation of the adjoint model is for example a difficult step in the variational algorithms. To get rid of these difficulties, we have very recently introduced a new algorithm, based on the nudging technique.

4.1   The Nudging Algorithm

The standard nudging algorithm consists in adding to the state equations a feedback term, which is proportional to the difference between the observation and its equivalent quantity computed by the resolution of the state equations. The model appears then as a weak constraint, and the nudging term forces the state variables to fit as well as possible to the observations.

Let us remind the model

$$\begin{cases} \dfrac{dX}{dt} = F(X, U), & 0 < t < T, \\ X(0) = V. \end{cases} \tag{18}$$

We still suppose that we have an observation $X_{obs}(t)$ of the state variable $X(t)$. The nudging algorithm simply gives

$$\begin{cases} \dfrac{dX}{dt} = F(X, U) + K(X_{obs} - CX), & 0 < t < T, \\ X(0) = V, \end{cases} \tag{19}$$

where C is still the observation operator, and K is the nudging matrix. It is quite easy to understand that if K is large enough, then the state vector transposed into the observation space (through the observation operator) $CX(t)$ will tend towards the observation vector $X_{obs}(t)$. In the linear case (where F and C are linear operators), the forward nudging method is nothing else than the Luenberger observer, also called asymptotic observer, where the operator K can be chosen so that the error goes to zero when time goes to infinity [44].

This algorithm was first used in meteorology [35], and then has been used with success in oceanography [63] and applied to a mesoscale model of the atmosphere [57]. Many results have also been carried out on the optimal determination of the nudging coefficients K [56, 64, 65].

The nudging algorithm is usually considered as a sequential data assimilation method. If one solves Eq. (19) with a numerical scheme, then it is equivalent with the following algorithm:

$$\begin{cases} X_n^f = X_{n-1} + dt \times F(X_{n-1}, U), \\ X_n = X_n^f + K_n(X_{obs}(t_n) - C_n X_n^f), \end{cases} \tag{20}$$

which is exactly the Kalman filter's algorithm. Then, if at any time the nudging matrix K is set in an optimal way, it is quite easy to see that K will be exactly the Kalman gain matrix. It is also possible to consider suboptimal K matrices, that still correct all variables, and not only the observed ones [10].

4.2 Backward Nudging

The backward nudging algorithm consists in solving the state equations of the model backwards in time, starting from the observation of the state of the system at the final instant. A nudging term, with the opposite sign compared to the standard nudging algorithm, is added to the state equations, and the final obtained state is in fact the initial state of the system [4, 7].

We now assume that we have a final condition in (18) instead of an initial condition. This leads to the following backward equation

$$\begin{cases} \dfrac{d\tilde{X}}{dt} = F(\tilde{X}, U), \quad T > t > 0, \\ \tilde{X}(T) = \tilde{V}. \end{cases} \tag{21}$$

If we apply nudging to this backward model with the opposite sign of the feedback term (in order to have a well posed problem), we obtain

$$\begin{cases} \dfrac{d\tilde{X}}{dt} = F(\tilde{X}, U) - K(X_{obs} - C\tilde{X}), \quad T > t > 0, \\ \tilde{X}(T) = \tilde{V}. \end{cases} \tag{22}$$

Once again, it is easy to see that if K is large enough, the vector $CX(t)$ will tend (through the observation operator) towards the observation vector $X_{obs}(t)$.

4.3 The BFN Algorithm

The back and forth nudging (BFN) algorithm consists in solving first the forward (standard) nudging equation, and then the direct system backwards in time with a feedback term. After resolution of this backward equation, one obtains an estimate of the initial state of the system. We repeat these forward and backward resolutions with the feedback terms until convergence of the algorithm [7].

The BFN algorithm is then the following:

$$\begin{cases} \dfrac{dX_k}{dt} = F(X_k, U) + K(X_{obs} - CX_k), \\ X_k(0) = \tilde{X}_{k-1}(0), \end{cases}$$

$$\begin{cases} \dfrac{d\tilde{X}_k}{dt} = F(\tilde{X}_k, U) - K'(X_{obs} - C\tilde{X}_k), \\ \tilde{X}_k(T) = X_k(T), \end{cases} \tag{23}$$

with $\tilde{X}_{-1}(0) = V$. Then, $X_0(0) = V$, and a resolution of the direct model gives $X_0(T)$ and hence $\tilde{X}_0(T)$. A resolution of the backward model provides $\tilde{X}_0(0)$, which is equal to $X_1(0)$, and so on.

This algorithm can be compared to the 4D-Var algorithm, which also consists in a sequence of forward and backward resolutions. In the BFN algorithm, even for nonlinear problems, it is useless to linearize the system and the backward system is not the adjoint equation but the direct system, with an extra feedback term that stabilizes the resolution of this ill-posed backward resolution.

The BFN algorithm has been tested successfully for the system of Lorenz equations, Burgers equation and a quasi-geostrophic ocean model in [8], for a shallow-water model in [5] and compared with a variational approach for all these models. It has been used to assimilate the wind data in a mesoscale model [14] and for the reconstruction of quantum states in [40].

4.4 DBFN: Diffusive Back and Forth Nudging Algorithm

In the framework of oceanographic and meteorological problems, there is usually no diffusion in the model equations. However, the numerical equations that are solved contain some diffusion terms in order to both stabilize the numerical integration (or the numerical scheme is set to be slightly diffusive) and model some subscale turbulence processes. We can then separate the diffusion term from the rest of the model terms, and assume that the partial differential equations read:

$$\partial_t X = F(X) + \nu \Delta X, \quad 0 < t < T, \tag{24}$$

where F has no diffusive terms, ν is the diffusion coefficient, and we assume that the diffusion is a standard second-order Laplacian (note that it could be a fourth or sixth order derivative in some oceanographic models, but for clarity, we assume here that it is a Laplacian operator).

We introduce the D-BFN algorithm in this framework, for $k \geq 1$:

$$\begin{cases} \partial_t X_k = F(X_k) + \nu \Delta X_k + K(X_{obs} - C(X_k)), \\ X_k(0) = \tilde{X}_{k-1}(0), \quad 0 < t < T, \end{cases} \quad \begin{cases} \partial_t \tilde{X}_k = F(\tilde{X}_k) - \nu \Delta \tilde{X}_k - K'(X_{obs} - C(\tilde{X}_k)), \\ \tilde{X}_k(T) = X_k(T), \quad T > t > 0. \end{cases} \tag{25}$$

It is straightforward to see that the backward equation can be rewritten, using $t' = T - t$:

$$\partial_{t'} \tilde{X}_k = -F(\tilde{X}_k) + \nu \Delta \tilde{X}_k + K'(X_{obs} - C(\tilde{X}_k)), \quad \tilde{X}_k(t' = 0) = X_k(T), \tag{26}$$

where $\tilde{X}$ is evaluated at time t'. Then the backward equation is well-posed, with an initial condition and the same diffusion operator as in the forward equation. The diffusion term both takes into account the subscale processes and stabilizes the numerical backward integrations, and the feedback term still controls the trajectory with the observations.

The main interest of this new algorithm is that for many geophysical applications, the non diffusive part of the model is reversible, and the backward model is then stable. Moreover, the forward and backward equations are now consistent in the sense that they will be both diffusive in the same way (as if the numerical schemes were the same in forward and backward integrations), and only the non-diffusive part of the physical model is solved backwards. Note that in this case, it is reasonable to set $K' = K$.

The DBFN algorithm has been tested successfully for a linear transport equation in [9], for non-linear Burgers equation in [6], and for full primitive equations in [53].

4.5   Theoretical Considerations

The convergence of the BFN algorithm has been proved by Auroux and Blum in [7] for linear systems of ordinary differential equations and full observations, by Ramdani et al. [51] for reversible linear partial differential equations (wave and Schrödinger equations), by Donovan et al. [20] for the reconstruction of quantum states. In [11], the authors consider the BFN algorithm on transport equations. They show that for non viscous equations (both linear transport and Burgers), the convergence of the algorithm holds under observability conditions. Convergence can also be proven for viscous linear transport equations under some strong hypothesis, but not for viscous Burgers' equation. Moreover, the authors show that the convergence rate is always exponential in time [11]. In [9], the authors prove the theoretical convergence of DBFN algorithm for linear transport equations.

Data Assimilation is the ensemble of techniques combining the mathematical information provided by the equations of the model and the physical information given by the observations in order to retrieve the state of a flow. In order to show that both BFN and DBFN algorithms achieve this double objective, let us give a formal explanation of the way these algorithms proceed.

If $K' = K$ and the forward and backward limit trajectory are equal, i.e. $\tilde{X}_\infty = X_\infty$, then taking the sum of the two equations in (23) shows that the limit trajectory X_∞ satisfies the model Eq. (18) (including possible model viscosity). Moreover, the difference between the two equations in (23) shows that the limit trajectory is solution of the following equation:

$$K(X_{obs} - C(X_\infty)) = 0. \tag{27}$$

Equation (27) shows that the limit trajectory perfectly fits the observations (through the observation operator, and the gain matrix).

In a similar way, for the DBFN algorithm, taking the sum of the two equations in (25) shows that the limit trajectory X_∞ satisfies the model equations without diffusion:

$$\partial_t X_\infty = F(X_\infty) \tag{28}$$

while taking the difference between the two same equations shows that X_∞ satisfies the Poisson equation:

$$\Delta X_\infty = -\frac{K}{\nu}(X_{obs} - C(X_\infty)) \tag{29}$$

which represents a smoothing process on the observations for which the degree of smoothness is given by the ratio $\frac{\nu}{K}$ [9]. Equation (29) corresponds, in the case where C is a matrix and $K = kC^T R^{-1}$, to the Euler equation of the minimization of the following cost function

$$J(X) = k\langle R^{-1}(X_{obs} - CX), (X_{obs} - CX)\rangle + \nu \int_\Omega \|\nabla X\|^2 \tag{30}$$

where the first term represents the quadratic difference to the observations and the second one is a first order Tikhonov regularisation term over the domain of resolution Ω. The vector X_∞, solution of (29), is the point where the minimum of this cost function is reached. This is a nice increment to the BFN algorithm, in which the limit trajectory fits the observations, while in the DBFN algorithm, the limit trajectory is the result of a smoothing process on the observations (which are often very noisy).

It is shown in the next section (numerical results) that at convergence the forward and backward trajectories are very close, which justifies this qualitative justification of the algorithm.

5 Numerical Results on a 2D Shallow-Water Model

5.1 Description of the Model

The shallow water model (or Saint-Venant's equations) is a basic model, representing quite well the temporal evolution of geophysical flows. This model is usually considered for simple numerical experiments in oceanography, meteorology or hydrology. The shallow water equations are a set of three equations, describing the evolution of a two-dimensional horizontal flow. These equations are derived from a vertical integration of the three-dimensional fields, assuming the hydrostatic approximation, i.e. neglecting the vertical acceleration. There are several ways to write the shallow water equations, considering either the geopotential or height or pressure variables. We consider here the following configuration:

$$
\begin{cases}
\partial_t u - (f + \zeta)v + \partial_x B = \dfrac{\tau}{\rho_0 h} - ru + \nu \Delta u, \\[2mm]
\partial_t v + (f + \zeta)u + \partial_y B = \dfrac{\tau}{\rho_0 h} - rv + \nu \Delta v, \\[2mm]
\partial_t h + \partial_x(hu) + \partial_y(hv) = 0,
\end{cases}
\tag{31}
$$

where the unknowns are u and v the horizontal components of the velocity, and h the geopotential height (see e.g. [1, 21]). The initial condition $(u(0), v(0), h(0))$ and no-slip lateral boundary conditions complete the system. The other parameters are the following:

- $\zeta = \partial_x v - \partial_y u$ is the relative vorticity;
- $B = g^* h + \dfrac{1}{2}(u^2 + v^2)$ is the Bernoulli potential;
- g^* is the reduced gravity;
- $f = f_0 + \beta y$ is the Coriolis parameter (in the β-plane approximation);
- $\tau = (\tau_x, \tau_y)$ is the forcing term of the model (e.g. the wind stress);
- ρ_0 is the water density; r is the friction coefficient; ν is the viscosity (or dissipation) coefficient.

We consider a numerical configuration in which the domain is a square of 2000×2000 km, with a rigid boundary, and no-slip boundary conditions. The time step is 1800 s (half an hour), and we consider an assimilation period $[0; T]$ of 720 time steps (i.e. 15 days). The forecast period is $[T; 4T]$, corresponding to 45 prediction

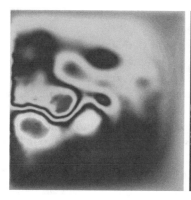

Fig. 1 Sea surface height h (*left*, in meters) and longitudinal velocity u (*right*, in ms^{-1}) of the ocean at a reference state. The ranges are the following: for h, from 250 m (*blue*) to 700 m (*red*), 500 being between *green* and *magenta*; for u, from -0.55 (*blue*) to 1 m per second (*red*), 0 being in *purple*

days. The spatial resolution is 25 km. The wind forcing is chosen constant in time, and set to a sine function which induces a standard double gyre circulation. This numerical model has been developed by the MOISE research team of INRIA Rhône-Alpes [21].

We briefly describe the numerical schemes used for the resolution of Eqs. (31) and we refer to [21] for more details. We consider a leap-frog method for time discretization of Eqs. (31), controlled by an Asselin time filter [3]. The equations are then discretized on an Arakawa C grid [2], with $N \times N$ points ($N = 81$ in our experiments): the velocity components u and v are defined at the center of the edges, and the height is defined at the center of the grid cells. Then, the vorticity and Bernoulli potential are computed at the nodes and center of the cells respectively.

The spin-up phase lasts nearly 6 years, starting from $u = v = 0$ and $h = 500$ meters, after which the model simulates a double-gyre wind-driven oceanic circulation. This approximate model reproduces quite well the surface circulation at mid-latitudes, including the jet stream and ocean boundary currents. In our experiments, the water depth varies from roughly 265–690 m, its mean being 500 m, and the maximum velocity (in the jet stream) is roughly 1.1 ms^{-1}, the mean velocity being 0.1 ms^{-1}. Figure 1 shows the height h and longitudinal velocity u at the reference state (true initial condition of the data assimilation period).

5.2 Experimental Setup

In all the following numerical experiments, we will consider twin experiments: a reference initial condition is set, and some data are extracted from the corresponding trajectory (see the model Eqs. (31) with $v = 0$). These data are then noised for some experiments (and not for some others), and provided as observations to the 4D-VAR, BFN and DBFN data assimilation algorithms.

We assume that some observations h_{obs} of only the height h are available, every n_t time steps and every n_x grid points (in both longitudinal and transversal directions). We can then easily define an observation operator C, from the model space to the observation space, as an extraction of one every n_x values of the variable h in each direction. This operator is clearly linear, and allows us to compare a model solution with the observations. Unless some other values are given, the considered values of n_x and n_t are respectively 5 and 24. Unless said otherwise (for instance in Sect. 5.6), the observations are unnoisy.

In such a configuration, the model space (state variables (u, v, h)) is of dimension 19,683, and the observation space (data variable h_{obs}) is of dimension 289. The corresponding total number of observations all over the assimilation period is 8959.

5.3 Convergence of DBFN

Figure 2 shows the relative RMS (root mean square) difference between the height of the DBFN iterates and the true height, versus time, for the 10 first iterations. The RMS error is computed in the following way:

$$Relative\ RMS = \frac{\|h_{DBFN} - h_{true}\|}{\|h_{true}\|}, \tag{32}$$

where we use the standard L^2 norm over the space domain.

The height of the background state (used for the initialization of the algorithm) has a relative error of 45 % with the true height at initial time. The error decreases during the first forward resolution, down to 28 %, and again during the first backward iteration, reaching approximately 22 % after one DBFN iteration. After 5 iterations, the algorithm almost converged, and the relative RMS error is close to 10 % (slightly more at the beginning of the assimilation window, and slightly less at the end).

In the same figure, one can see the relative RMS errors for the two other variables of the model, the longitudinal and transversal components of the velocity. The error on u decreases from 61 % to less than 30 % in 5 iterations, and the error on v decreases from more than 80 % to approximately 40 %. So we can consider that in 5 (or 6) iterations, the DBFN algorithm converged. Note that there are no correction terms in the DBFN velocity equations, as there are no observations on the velocity, but the model coupling between the three variables allows the DBFN to correct all variables and not only the sea surface height.

Figure 3 shows the evolution of the relative RMS error of the DBFN solution, as a function of time. Between day 0 and day 15, the assimilation window, one can see the DBFN iterations with the decrease of the error during forward and backward iterations. Then, once the algorithm reaches convergence, we use the state identified

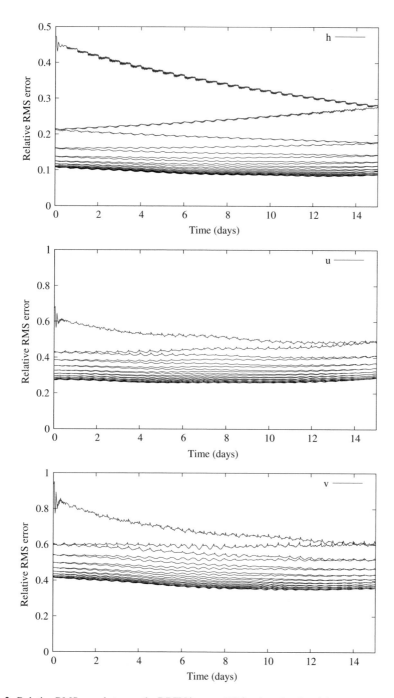

Fig. 2 Relative RMS error between the DBFN iterates (10 first iterations) and the true state, versus time, for the sea surface height h, and the velocity u and v

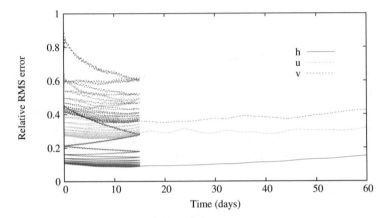

Fig. 3 Evolution of relative RMS error of the DBFN solution during the iterations (15 first days, corresponding to the assimilation window), and the forecast (from day 15 to day 60, corresponding to the forecast window)

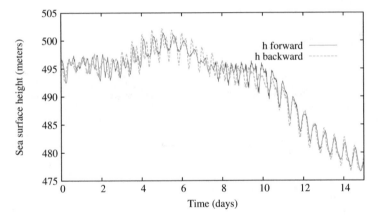

Fig. 4 Trajectory of the sea surface height at one particular point in the space domain during one back and forth iteration

by the DBFN algorithm as the initial condition for a forecast experiment: the model, without any feedback term, is used to propagate the solution in time, and predict the evolution of the solution.

Note that on the first 15 days, Fig. 3 shows the same results as in the three plots of Fig. 2. The interesting part is on the next 45 days, corresponding to the forecast period. The error remains quite stable and the final forecasted state has less than 20 % error on the sea surface height, and approximately 40 % on the velocity.

Figure 4 shows the evolution of the sea surface height at a given point in the domain (close to the main jet in the west part of the domain) during one back and forth iteration (iteration 5, when the DBFN converged). One can see that forward and backward solutions are very close, and this confirms that the algorithm converged, as the new initial value (at time 0) after the backward iteration is similar to the previous initial value before the forward iteration.

5.4 Comparison Between BFN, DBFN and 4DVAR

Figure 5 shows the sea surface height h of several initial conditions (at time 0): on the top left, the initial condition identified by the BFN after 5 iterations (converged); on the top right, the initial condition identified after 5 iterations of DBFN (converged); on the bottom left, the initial condition identified by the 4DVAR after 50 iterations (converged); and on the bottom right, the true initial condition.

Note that we looked at the solution of the 4DVAR after 5 iterations. The computational costs of one (D)BFN iteration and of one 4DVAR iteration are more or less comparable, as they both consist in one resolution of the direct model, and one resolution of either the backward or the adjoint model. But the identified solution after 5 iterations of 4DVAR is not good at all and still too close to the

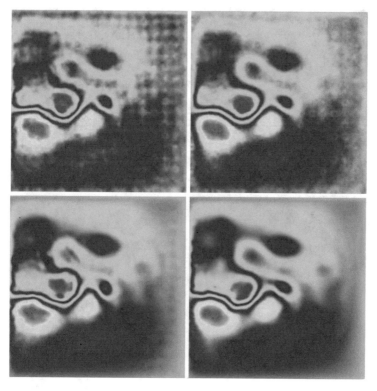

Fig. 5 Initial condition (sea surface height) identified by: BFN (5 iterations, converged), DBFN (5 iterations, converged), 4DVAR (50 iterations, converged), true solution

background state. So we decided to show only the solution after 50 iterations, when the algorithm converged.

We can see on Fig. 5 that the DBFN solution looks nicer than the BFN solution. In the BFN algorithm, the identified initial condition is the result of the last backward integration, where diffusion has the wrong sign. This partly explains why the BFN solution is not so smooth. The DBFN solution is better than the BFN one, as the diffusion helps for a better filtering of noise in the background state (and also a better filtering of the feedback term that is added at only some locations). All solutions show a pretty good agreement with the true state, with a good location of the Gulf stream and of the main vortices, but (D)BFN solutions are obtained in a much smaller time, as they require 10 times less iterations.

Figure 6 shows similar results as in Fig. 5, with one component u of the velocity instead of the sea surface height h. The color scale goes from -0.6 (blue) to $1.3\,\mathrm{ms}^{-1}$ (red). We can see that BFN, DBFN and 4DVAR identified solutions are close to the true state. But they all present some small oscillations and waves outside the Gulf stream zone. However, these small effects are negligible and do not induce big differences in the result. As previously said, it is noticeable that the velocity is

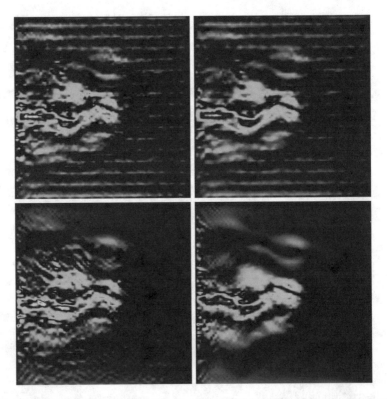

Fig. 6 Initial condition (longitudinal velocity u) identified by: BFN (5 iterations, converged), DBFN (5 iterations, converged), 4DVAR (50 iterations, converged), true solution

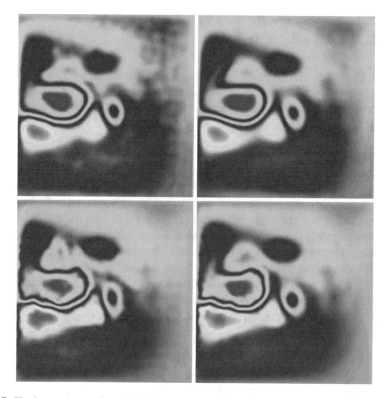

Fig. 7 Final state (sea surface height at the end of the forecast period) identified by: BFN (5 iterations, converged), DBFN (5 iterations, converged), 4DVAR (50 iterations, converged), true solution

quite well identified while there are no observations on velocity, only on sea surface height.

Finally, Fig. 7 shows similar results as in the previous figures, for the sea surface height h, at the end of the forecast period (60 days). From the identified initial conditions (Figs. 5 and 6), we used the model to propagate the solution in time, without any assimilation, up to 60 days.

We can see that all solutions are still close to the true state. The DBFN solution looks smoother than the BFN one (as the identified initial condition was also smoother). The DBFN solution has been obtained with a much smaller computational cost (10 times less) than the 4DVAR solution, as it required only 5 iterations instead of 50. As for the initial condition, if we stop the 4DVAR after 5 iterations, the solution is not good at all.

We can conclude that the DBFN gives smoother solutions than the BFN, which is interesting for noise filtering. This is indeed due to the reversed diffusion in the backward resolutions. And the DBFN gives similar results to the 4DVAR, but much faster as it requires much less iterations to achieve convergence.

5.5 Sensitivity of the BFN and DBFN with Respect to the Diffusion Coefficient

Table 1 gives the relative RMS error of the solution identified by the BFN and DBFN, for several viscosity coefficients. For each case, the error is measured at both time $t = 0$ (initial condition identified by the algorithm at convergence), and at time $t = 60$ days (corresponding final forecast obtained from the integration of the model).

When the diffusion is very small, BFN and DBFN give very similar results. As the diffusion has a very limited role in this case, BFN and DBFN algorithms are very close (and equal in the limit case where $v = 0$). When the diffusion increases, the DBFN algorithms gives better results, particularly on the velocity components. This is due to the fact that the DBFN has a better filtering role than the BFN algorithm, as there is also diffusion (with the good sign) in the backward integrations. As we only have observations on a sparse grid, and as the feedback term is added only at observation locations, the filtering effect helps the feedback to be smoothed over neighboring grid points. Finally, if the diffusion is very high, both methods give degraded results: the BFN has strong instabilities in the backward resolutions, and the DBFN has a too strong diffusive effect. However, the errors on the forecast remain very reasonable.

Table 1 Relative RMS error (in %) of the solution identified by the (D)BFN for several viscosity coefficients (in $m^2.s^{-1}$): error on the identified initial condition after 5 iterations (converged), and on the corresponding final forecast at time $t = 60$ days

		h init	u init	v init	h forecast	u forecast	v forecast
$v = 0.5$	BFN	14.9	45.6	65.1	22.6	46.7	66.5
	DBFN	14.7	44.6	63.6	21.9	45.5	64.0
$v = 1.5$	BFN	14.9	45.6	65.2	22.5	46.4	65.8
	DBFN	14.3	42.8	61.1	20.9	43.6	59.6
$v = 5$	BFN	14.9	45.7	65.4	22.4	45.6	64.1
	DBFN	13.0	38.8	54.6	18.5	37.3	48.7
$v = 15$	BFN	14.9	46.0	65.9	21.9	43.6	59.9
	DBFN	10.9	32.8	45.4	16.4	33.2	44.6
$v = 50$	BFN	15.0	47.7	68.3	21.0	39.2	52.3
	DBFN	10.8	27.1	41.6	15.1	31.9	42.4
$v = 150$	BFN	15.6	57.7	82.0	20.1	36.2	46.1
	DBFN	14.6	30.1	49.7	16.8	36.9	46.4

Table 2 Relative RMS error (in %) of the solution identified by the (D)BFN for several observation noise levels: error on the identified initial condition after 5 iterations (converged), and on the corresponding final forecast at time $t = 60$ days

		h init	u init	v init	h forecast	u forecast	v forecast
Noise = 5 %	BFN	15.3	49.6	69.6	20.7	38.6	51.5
	DBFN	11.2	28.0	42.6	15.2	32.0	42.8
Noise = 10 %	BFN	16.1	53.0	74.4	22.0	39.7	51.5
	DBFN	12.3	30.6	46.0	15.4	32.2	43.4
Noise = 20 %	BFN	19.3	64.1	91.3	23.9	42.7	55.8
	DBFN	15.0	39.7	58.2	15.7	33.2	45.0
Noise = 40 %	BFN	29.6	98.1	136.5	31.3	58.2	74.5
	DBFN	23.3	65.6	93.7	17.3	36.8	50.1

5.6 Sensitivity of the BFN and DBFN with Respect to Observation Noise

In this section, we now consider noisy observations. Table 2 gives the relative RMS error of the solution identified by the BFN and DBFN, for several observation noise levels. In this experiment, we set the diffusion coefficient to $5\,\mathrm{m}^2.\mathrm{s}^{-1}$. We added Gaussian white noise to the observations, with several relative levels. For each case, the error is measured at both time $t = 0$ (initial condition identified by the algorithm at convergence), and at time $t = 60$ days (corresponding final forecast obtained from the integration of the model).

When the noise level increases, of course all initial conditions identified are degraded, but the DBFN is less sensitive to noise on the observations, particularly on the velocity components. For all noise levels, the DBFN gives better results than the BFN on all components. If we look at the results on the final forecast state, we can see that the DBFN is much less sensitive to observation noise than the BFN, and it gives much better results for high levels (though realistic) of noise.

6 Numerical Results on a 3D Primitive Equation Ocean Model

6.1 Description of the Model

In this section the BFN and DBFN are tested on a Primitive Equation (PE) ocean model. The ocean model used in this study is the ocean component of NEMO (Nucleus for European Modeling of the Ocean; [46]). This model is able to represent a wide range of ocean motions, from basin scale up to regional scale. Currently, it has been used in operational mode at the French Mercator Océan project (http://www.mercator-ocean.fr).

The model solves six prognostic equations for zonal velocity (u) and meridional velocity (v), pressure (P), sea surface height (η), temperature (T), salinity (S) and finally a state equation which connects the mass field to the dynamical field. Two physically-based hypotheses are made: Boussinesq and incompressible ocean, and the hydrostatic approximation.

The conservation of momentum is expressed as the Navier-Stokes equations for a fluid element located at (x, y, z) on the surface of our rotating planet and moving at velocity (u, v, w) relative to that surface:

$$\frac{DU}{Dt} = -\frac{\nabla P}{\rho} + g - 2\Omega \times U + D + F, \tag{33}$$

where t is the time, $U = U_h + w\mathbf{k}$ is the velocity vector composed by its horizontal components $U_h = (u, v)$ and vertical component $w\mathbf{k}$, where $\mathbf{k}$ is the upward unit vector perpendicular to the earth surface, g is the gravitational acceleration, Ω is the rotational vector, ρ is the density, and D and F are the dissipation and forcing terms. The operator $\frac{D}{Dt}$ is the total derivative which includes time and local variations. It is given by:

$$\frac{D}{Dt} = \frac{\partial}{\partial t} + U.\nabla.$$

The conservation of mass is given by the continuity equation:

$$\frac{\partial \rho}{\partial t} + \nabla (\rho U) = 0 . \tag{34}$$

The Boussinesq approximation considers that density variations are much smaller than the mean density

$$\frac{\delta \rho}{\rho} \ll 1 .$$

Therefore, the ocean density is considered constant, i.e. a mean density ρ_0 replaces ρ in the equations, except in the buoyancy term for which the density is multiplied by the gravity. Considering the aspect ratio $\frac{\tilde{H}}{\tilde{L}} \ll 1$, where $\tilde{L}$ is the horizontal scale and $\tilde{H}$ is the vertical scale, the equation for the evolution of w is resumed to the hydrostatic equilibrium:

$$\frac{\partial P}{\partial z} = -\rho g \tag{35}$$

With this approximation the acceleration term, $\dfrac{\partial w}{\partial t}$, is neglected implying a misrepresentation of gravitational flows as well as of vertical convection processes. Indeed, the vertical component of the velocity field is a diagnostic variable calculated thanks to the continuity equation under the Boussinesq approximation:

$$\frac{\partial w}{\partial z} = -\nabla_h.U, \tag{36}$$

where ∇_h is the restriction of the gradient operator to the horizontal plane.

The equation for the evolution of the free surface, i.e. for the sea surface height η, is derived from the surface kinetic boundary condition and may be written as:

$$\frac{\partial \eta}{\partial t} = -\nabla_h(D\bar{U}_h)$$
$$\bar{U}_h = \frac{1}{D} \int_{-D}^{0} U_h dz, \tag{37}$$

where D is the water depth.

The equations for the conservation of salt and potential temperature are derived from the first thermodynamical law. The final equations can be written as:

$$\frac{\partial T}{\partial t} = -\nabla (UT) + D^T + F^T \tag{38}$$

$$\frac{\partial S}{\partial t} = -\nabla (US) + D^S + F^S . \tag{39}$$

The system is closed by a state equation linking density, temperature, salinity and pressure:

$$\rho = \rho(T, S, P) . \tag{40}$$

The system composed by Eqs. (33),(37),(38), (39) and (40) are discretized on a spherical mesh using a traditional centered second-order finite difference approximation. The non-diffusive terms evolves in time using a leap-frog method, controlled by an Asselin time filter [3], and the diffusion terms are discretized using a forward Euler scheme, for the horizontal diffusion, and an implicity backward Euler, for the vertical diffusion terms, in order to assure numerical stability.

Special attention has been given to the homogeneity of the solution in the three space directions. The arrangement of variables is the same in all directions. It consists of cells centered on scalar points (T, S, P, ρ) with vector points (u, v, w) defined in the center of each face of the cells. This is the well-known C grid in Arakawa's classification generalized to the three dimensional space. The relative and planetary vorticity, ζ and f, are defined in the center of each vertical edge and the barotropic stream function ϕ is defined at the horizontal points overlying the ζ

and f points. More details on the model formulation and on the numerical methods are given by [46].

6.2 Model Configuration

NEMO model is configured to simulate the double gyre circulation as the SW model used in the Sect. 5. Two experiments using different horizontal resolution are configured: one at 9.25 km resolution, which is used as representing the true ocean state, and another one with 25.5 km resolution that will assimilate the observations from the higher resolution model. For both configurations 11 vertical levels are considered with resolution ranging from 100 m near the upper surface up to 500 m near the bottom. The bottom topography is flat and the lateral boundaries are closed and frictionless. The model depth is set to $H = 5000$ m, in this case the variable h described in the Sect. 5 corresponds to $h = H + \eta$. The only forcing term considered is a constant wind stress of the form $\tau = \left(\tau_0 cos \left(\frac{2\pi(y - y_0)}{L} \right), 0 \right)$, where $L = 2000$ km and $\tau_0 = 0.1$ N/m^2. The diffusion and viscosity terms D, D^T and D^S in the Eqs. (33), (38) and (39) are decomposed into a horizontal component, which is modeled by a bilaplacian operator $\left(D_h = v_h \nabla_h^4 \right)$, and a vertical component, which is modeled by a laplacian operator $\left(D_v = v_v \frac{\partial^2}{\partial z^2} \right)$. They all use constant coefficients in time and space. Table 3 summarizes the characteristics of each experiment in terms of number of grid points, time step and horizontal and vertical diffusion/viscosity coefficients.

The initial condition consists of a homogeneous salinity field of 35 psu and a temperature field created to provide a stratification which has a first baroclinic deformation radius of 44.7 km. Velocity and pressure fields are initially set to zero. For both experiments the model was integrated for 70 years, in order to reach the statistical steady state. The final condition of the simulation at 25.5 km resolution is used as the first guess in the assimilation experiments. The 9.25 km resolution experiment is further integrated for 1 year to generate the true trajectory.

Table 3 Characteristics of each experiment in terms of horizontal spatial resolution, number of grid points (N_x, N_y, N_z), time step and horizontal (v_h) and vertical (v_v) diffusion/viscosity coefficients

Resolution	N_x	N_y	N_z	time step (s)	v_h (m^4/s)	v_v (m^2/s)
9.25 km	363	243	11	300	-8×10^9	1.2×10^{-5}
25.5 km	121	81	11	900	-8×10^{10}	1.2×10^{-5}

6.3 Experimental Setup

6.3.1 Model Aspects

The assimilation period $[0; T]$ is made of 960 time steps (i.e. 10 days). Each dataset (see Sect. 6.3.2) is assimilated with the BFN and DBFN using a reduced horizontal diffusion coefficient, $v_h = -8 \times 10^8 \, \text{m}^4/\text{s}$, which is different from the diffusion coefficient used to generate the background field (see Table 3). The vertical diffusion coefficient is the same as in Table 3.

6.3.2 Observation Network

The observation network is composed by observations of the Sea Surface Height (η) sampled in order to mimic the spatial sample density of the Jason-1 satellite, but with higher time sampling frequency (here we considered the satellite period as 1 day in contrast to the 10 day period of Jason-1), complete daily fields of Sea Surface Temperature (SST), which may be seen as the merged products of SST derived from satellite observations, and vertical profiles of temperature and salinity available every $3° \times 3°$ and every day, similar to the ARGO buoys sampling. This observation network is quite close to a real ocean observation system in terms of number of available observations within the given model domain.

The true state, used to generate the observations, comes from a much higher resolution experiment ($\Delta x = 9.25 \, \text{km}$). As a consequence, the observations contain two types of error: representation errors that simulate the fact that the observations are sampled from a system containing process and scales that are not modeled by the model used in the assimilation step, and white Gaussian errors that simulate instrumental errors.

For each DBFN and BFN configuration two datasets are assimilated: one without white noise and another one with a noise level of 100 %.

6.4 Convergence of DBFN

Our reference experiment assimilates observations without noise with the DBFN. Figure 8 shows the evolution of the relative error during the forward and backward integrations for the SSH (η) and for the zonal velocity, which is a non observed variable. The DBFN needs 108 iterations to converge, which is much larger than the number of iterations needed by the DBFN to converge in the SW experiments. This is due to the fact that the observation information takes time to be propagated downwards to the deep ocean, and for the temperature increments, calculated from the vertical profiles, to be smoothed.

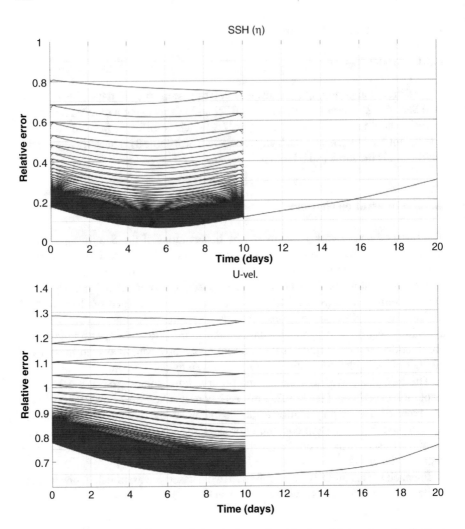

Fig. 8 Evolution of the relative error during the forward and backward integrations for the (*top*) SSH and (*bottom*) zonal velocity for the experiment assimilating observations without noise with the DBFN. *Red curves* represent the forward integration and *blue curves* represent the backward integration. The *red curve* for the days 10–20 represents a forecast initialized with the initial condition identified by the DBFN

6.5   *Comparison Between DBFN and BFN*

Table 4 summarizes the relative error on the initial condition obtained for both algorithms in the case of the assimilation of observations without random noise and with 100 % of noise. The DBFN produces the best results for both cases: assimilating observations with and without noise. Indeed, the DBFN performance

Table 4 Summary of the relative error on the initial condition obtained after 108 iterations, which is the number of iterations needed by the DBFN-0 to converge, for both algorithms in the case of the assimilation of observations without random noise (BFN-0 and DBFN-0) and with 100 % of noise (BFN-1 and DBFN-1)

	SSH	u	v	T	S
BFN-0	0.1790	0.8226	0.7130	0.1195	0.0143
DBFN-0	0.1754	0.7790	0.6440	0.1193	0.0143
BFN-1	0.1926	0.8928	0.7932	0.1195	0.0143
DBFN-1	0.1872	0.8053	0.6710	0.1194	0.0143

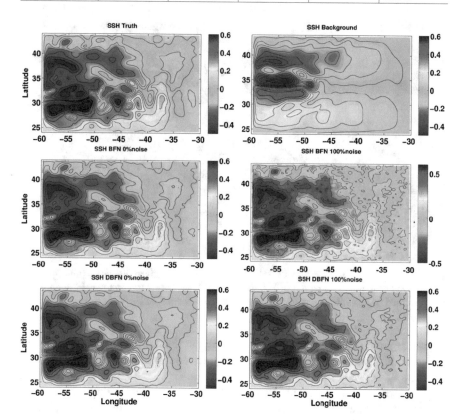

Fig. 9 SSH fields for the true state (*top left*) and the first guess (*top right*), and the analyzed state produced by the BFN (*middle panels*) and DBFN (*bottom panels*) assimilating non perturbed (*left panels*) and perturbed (*right panels*) observations

is improved relative to the BFN performance when the observations are noisy. This is the same result observed for the SW model and is related to the diffusive aspect of the algorithm which helps the method to filter out the observation noise.

The Figs. 9 and 10 show the SSH and surface zonal velocity fields for the first guess, the true state and the analyzed states produced by the BFN and DBFN assimilating perturbed and non perturbed observations. The first important remark

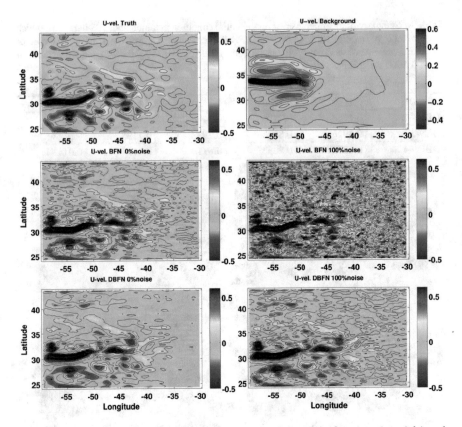

Fig. 10 Surface zonal velocity fields for the true state (*top left*) and the first guess (*top right*), and the analyzed state produced by the BFN (*middle panels*) and DBFN (*bottom panels*) assimilating non perturbed (*left panels*) and perturbed (*right panels*) observations

is that the front position in the true state is displaced southward with respect to the background. This is observed for the entire free run and is the effect of non-linear interactions between the small scales, not modeled by the 25.5 km resolution, and the large scale. All experiments were able to replace the front at the right position. Moreover, all experiments were able to reproduce the presence of eddies, especially the three eddies observed southward of the main front.

The results also show that even in the absence of noise in the observations, the velocity field produced by the BFN is quite noisy due to the diffusion in the backward integration. In the presence of noise, the obtained BFN solution has an increased level of noise. The same is also valid for the DBFN, but the observation noise has a much smaller impact in this case. In other words, the DBFN has a double advantage: first it ensures the stability of the backward integration, and second it is more efficient in filtering out the observation errors.

6.6   Concluding Remarks

In this section we have seen that the BFN and the DBFN can be used to assimilate oceanic observations into a PE ocean model. Indeed, for a PE ocean model, the DBFN should be preferably used due to its diffusive aspect that has a double role: stabilize the backward integration and filter out the noise on the observations.

7   Conclusion

The diffusive back and forth nudging algorithm has a very easy implementation, which does not impose any linearization of the model, which does not require neither the construction of an adjoint model, nor an optimization procedure as in the 4DVAR method. Moreover it is a very fast algorithm, which converges in less iterations than the 4DVAR. Therefore it has a lower computational and memory cost than the variational method. The progress of the DBFN, compared to the BFN method, is a better smoothing of the noise on the observations, which results from a more diffusive process (forward and backward). The distance from the reconstructed solution to the true one is also smaller with the DBFN algorithm than the one obtained with the BFN method. These characteristics have been observed, both on 2D shallow water model and on the 3D ocean primitive equation model.

The perspective is to achieve an optimization of the gain matrix, which can be the same in the forward and in the backward resolution, which is not the case in the BFN method. And the final goal is of course to apply this method to real data in order to confirm the conclusions obtained here on realistic models and with synthetic data corresponding to realistic observation systems.

Another interesting point would be to obtain a proof of the convergence of the method on these sophisticated non-linear geophysical models.

Acknowledgements The authors would like to thank Yann Brenier (CMLS, École Polytechnique), Emmanuel Cosme (MEOM, LGGE Grenoble) and Maëlle Nodet (MOISE, LJK Grenoble) for fruitful discussions. This work was supported by a CNRS-INSU LEFE-MANU grant.

References

1. Adcroft, A., Marshall, D.: How slippery are piecewise-constant coastlines in numerical ocean models? Tellus **50**(1), 95–108 (1998)
2. Arakawa, A., Lamb, V.: Computational design of the basic dynamical processes of the UCLA general circulation model. Methods Comput. Phys. **17**, 174–267 (1977)
3. Asselin, R.: Frequency filter for time integrations. Mon. Weather Rev. **100**, 487–490 (1972)
4. Auroux, D.: Étude de différentes méthodes d'assimilation de données pour l'environnement. PhD thesis, University of Nice Sophia-Antipolis, France (2003)

5. Auroux, D.: The back and forth nudging algorithm applied to a shallow water model, comparison and hybridization with the 4D-VAR. Int. J. Numer. Methods Fluids **61**(8), 911–929 (2009)
6. Auroux, D., Bansart, P., Blum, J.: An evolution of the Back and Forth Nudging for geophysical data assimila tion: application to burgers equation and comparisons. Inverse Probl. Sci. Eng. **21**(3), 399–419 (2013)
7. Auroux, D., Blum, J.: Back and forth nudging algorithm for data assimilation problems. C. R. Acad. Sci. Paris Ser. I **340**, 873–878 (2005)
8. Auroux, D., Blum, J.: A nudging-based data assimilation method for oceanographic problems: the back and forth nudging (BFN) algorithm. Nonlinear Proc. Geophys. **15**, 305–319 (2008)
9. Auroux, D., Blum, J., Nodet, M.: Diffusive Back and Forth Nudging algorithm for data assimilation. C. R. Acad. Sci. Paris Ser. I **349**(15–16), 849–854 (2011)
10. Auroux, D., Bonnabel, S.: Symmetry-based observers for some water-tank problems. IEEE Trans. Autom. Control **56**(5), 1046–1058 (2011)
11. Auroux, D., Nodet, M.: The Back and Forth nudging algorithm for data assimilation problems: theoretical results on transport equations. ESAIM Control Optim. Calc. Var. **18**(2), 318–342 (2012)
12. Bennett, A.F.: Inverse Modeling of the Ocean and Atmosphere. Cambridge University Press, Cambridge (2002)
13. Blum, J., Le Dimet, F.-X., Navon, I.M.: Data assimilation for geophysical fluids. In: Temam, R., Tribbia, J.J. (eds.) Computational Methods for the Atmosphere and the Oceans. Handbook of Numerical Analysis, vol. XIV, pp. 385–441. Elsevier, Amsterdam/London (2009)
14. Boilley, A., Mahfouf, J.-F.: Assimilation of low-level wind in a high resolution mesoscale model using the Back and Forth nudging algorithm. Tellus A **64**, 18697 (2012)
15. Broyden, C.G.: A new double-rank minimization algorithm. Not. Am. Math. Soc. **16**, 670 (1969)
16. Cane, M.A., Kaplan, A., Miller, R.N., Tang, B., Hackert, E.C., Busalacchi, A.J.: Mapping tropical Pacific Sea level: data assimilation via a reduced state Kalman filter. J. Geophys. Res. **101**(C10), 22599–22617 (1996)
17. Carrassi, A., Vannitsem, S.: Deterministic treatment of model error in geophysical data assimilation. In: Ancona, F., et al. (eds.) Mathematical Paradigms of Climate Science. Springer (2016)
18. Courtier, P., Talagrand, O.: Variational assimilation of meteorological observations with the adjoint equations Part 2. Numerical results. Q. J. R. Meteor. Soc. **113**, 1329–1347 (1987)
19. Le Dimet, F.-X., Talagrand, O.: Variational algorithms for analysis and assimilation of meteorogical observations: theoretical aspects. Tellus **38A**, 97–110 (1986)
20. Donovan, A., Mirrahimi, M., Rouchon, P.: Back and Forth nudging for quantum state reconstruction. In: 4th International Symposium Communications Control Signal Proceedings, Limassol, pp. 1–5, Mar 2010
21. Durbiano, S.: Vecteurs caractéristiques de modèles océaniques pour la réduction d'ordre en assimilation de données. PhD thesis, University of Grenoble I, France (2001)
22. Evensen, G.: Using the extended Kalman filter with a multilayer quasi-geostrophic ocean model. J. Geophys. Res. **97**, 17905–17924 (1992)
23. Evensen, G.: Sequential data assimilation with a nonlinear quasi-geostrophic model using Monte Carlo methods to forecast error statistics. J. Geophys. Res. **99**(C5), 10143–10162 (1994)
24. Evensen, G.: The ensemble Kalman filter: theoretical formulation and practical implementation. Ocean Dyn **53**, 343–367 (2003)
25. Evensen, G.: Data Assimilation: The Ensemble Kalman Filter. Springer, Berlin/Heidelberg (2009)
26. Fukumori, I.: Assimilation of topex sea level measurements with a reduced-gravity, shallow water model of the tropical Pacific Ocean. J. Geophys. Res. **100**(C12), 25027–25039 (1995)
27. Fukumori, I., Benveniste, J., Wunsch, C., Haidvogel, D.B.: Assimilation of sea surface topography into an ocean circulation model using a steady state smoother. J. Phys. Oceanogr. **23**, 1831–1855 (1993)

28. Gauthier, P., Courtier, P., Moll, P.: Assimilation of simulated wind lidar data with a Kalman filter. Mon. Weather Rev. **121**, 1803–1820 (1993)
29. Gelb, A.: Applied Optimal Estimation. MIT, Cambridge (1974)
30. Ghil, M.: Meteorological data assimilation for oceanographers. Part 1: description and theoretical framework. Dyn. Atmos. Oceans **13**, 171–218 (1989)
31. Ghil, M., Cohn, S.E., Dalcher, A.: Sequential estimation, data assimilation and initialization. In: Williamson, D. (ed.) The Interaction Between Objective Analysis and Initialization. Publication in Meteorology, vol. 127. McGill University, Montréal (1982)
32. Ghil, M., Manalotte-Rizzoli, P.: Data assimilation in meteorology and oceanography. Adv. Geophys. **23**, 141–265 (1991)
33. Gilbert, J.-Ch., Lemaréchal, C.: Some numerical experiments with variable storage quasi-Newton algorithms. Math. Prog. **45**, 407–435 (1989)
34. Gourdeau, L., Arnault, S., Ménard, Y., Merle, J.: Geosat sea-level assimilation in a tropical Atlantic model using Kalman filter. Ocean. Acta **15**, 567–574 (1992)
35. Hoke, J., Anthes, R.A.: The initialization of numerical models by a dynamic initialization technique. Mon. Weaver Rev. **104**, 1551–1556 (1976)
36. Holland, W.R.: The role of mesoscale eddies in the general circulation of the ocean. J. Phys. Ocean **8**(3), 363–392 (1978)
37. Houtekamer, P., Mitchell, H.: Data assimilation using an ensemble Kalman filter technique. Mon. Weather Rev. **126**, 796–811 (1998)
38. Jazwinski, A.H.: Stochastic Processes and Filtering Theory. Academic, New York (1970)
39. Kalnay, E.: Atmospheric Modeling, Data Assimilation and Predictability. Cambridge University Press, Cambridge/New York (2003)
40. Leghtas, Z., Mirrahimi, M., Rouchon, P.: Observer-based quantum state estimation by continuous weak measurement. In: American Control Conference (ACC), pp. 4334–4339 (2011)
41. Lewis, J.M., Derber, J.C.: The use of adjoint equations to solve a variational adjustment problem with convective constraints. Tellus **37A**, 309–322 (1985)
42. Lions, J.L.: Contrôle optimal de systèmes gouvernés par des équations aux dérivées partielles. Dunod (1968)
43. Liu, D.C., Nocedal, J.: On the limited memory BFGS method for large scale optimization. Math. Prog. **45**, 503–528 (1989)
44. Luenberger, D.: Observers for multivariable systems. IEEE Trans. Autom. Control **11**, 190–197 (1966)
45. Luong, B., Blum, J., Verron, J.: A variational method for the resolution of a data assimilation problem in oceanography. Inverse Probl. **14**, 979–997 (1998)
46. Madec, G.: NEMO ocean engine. Note du Pôle de modélisation, Institut Pierre-Simon Laplace (IPSL) No 27, France (2008)
47. Mohammadi, B., Pironneau, O.: Applied Shape Optimization for Fluids. Clarendon Press, Oxford (2001)
48. Moore, A.M.: Data assimilation in a quasigeostrophic open-ocean model of the Gulf-stream region using the adjoint model. J. Phys. Oceanogr. **21**, 398–427 (1991)
49. Nechaev, V., Yaremchuk, M.I.: Application of the adjoint technique to processing of a standard section data set: World Ocean circulation experiment section s4 along $67°s$ in the Pacific Ocean. J. Geophys. Res. **100**(C1), 865–879 (1994)
50. Pedlosky, J.: Geophysical Fluid Dynamics. Springer, New-York (1979)
51. Ramdani, K., Tucsnak, M., Weiss, G.: Recovering the initial state of an infinite-dimensional system using observers. Automatica **46**(10), 1616–1625 (2010)
52. Rostaing-Schmidt, N., Hassold, E.: Basic function representation of programs for automatic differentiation in the Odyssée system. In: Le Dimet, F.-X. (ed.) High Performance Computing in the Geosciences, pp. 207–222. Kluwer Academic (1994)
53. Ruggiero, G.A., Ourmières, Y., Cosme, E., Blum, J., Auroux, D., Verron, J.: Data assimilation experiments using diffusive Back-and-Forth nudging for the nemo ocean model. Nonlinear Process. Geophys. **22**(2), 233–248 (2015)

54. Schröter, J., Seiler, U., Wenzel, M.: Variational assimilation of geosat data into an eddy-resolving model of the Gulf stream area. J. Phys. Oceanogr. **23**, 925–953 (1993)
55. Sheinbaum, J., Anderson, D.L.T.: Variational assimilation of XBT data. Part 1. J. Phys. Oceanogr. **20**, 672–688 (1990)
56. Stauffer, D.R., Bao, J.W.: Optimal determination of nudging coefficients using the adjoint equations. Tellus A **45**, 358–369 (1993)
57. Stauffer, D.R., Seaman, N.L.: Use of four dimensional data assimilation in a limited area mesoscale model – Part 1: experiments with synoptic-scale data. Mon. Weather Rev. **118**, 1250–1277 (1990)
58. Talagrand, O.: Assimilation of observations, an introduction. J. Meteorol. Soc. Jpn. **75**(1B), 191–209 (1997)
59. Talagrand, O., Courtier, P.: Variational assimilation of meteorological observations with the adjoint vorticity equation. Part 1: theory. Q. J. R. Meteorol. Soc. **113**, 1311–1328 (1987)
60. Thacker, W.C., Long, R.B.: Fitting dynamics to data. J. Geophys. Res. **93**, 1227–1240 (1988)
61. Veersé, F., Auroux, D., Fisher, M.: Limited-memory BFGS diagonal preconditioners for a data assimilation problem in meteorology. Optim. Eng. **1**(3), 323–339 (2000)
62. Verron, J., Gourdeau, L., Pham, D.T., Murtugudde, R., Busalacchi, A.J.: An extended Kalman filter to assimilate satellite altimeter data into a non-linear numerical model of the tropical pacific ocean: method and validation. J. Geophys. Res. **104**, 5441–5458 (1999)
63. Verron, J., Holland, W.R.: Impact de données d'altimétrie satellitaire sur les simulations numériques des circulations générales océaniques aux latitudes moyennes. Ann. Geophys. **7**(1), 31–46 (1989)
64. Vidard, A., Le Dimet, F.-X., Piacentini, A.: Determination of optimal nudging coefficients. Tellus A **55**, 1–15 (2003)
65. Zou, X., Navon, I.M., Le Dimet, F.-X.: An optimal nudging data assimilation scheme using parameter estimation. Q. J. R. Meteorol. Soc. **118**, 1163–1186 (1992)

Deterministic Treatment of Model Error in Geophysical Data Assimilation

Alberto Carrassi and Stéphane Vannitsem

Abstract This chapter describes a novel approach for the treatment of model error in geophysical data assimilation. In this method, model error is treated as a deterministic process correlated in time. This allows for the derivation of the evolution equations for the relevant moments of the model error statistics required in data assimilation procedures, along with an approximation suitable for application to large numerical models typical of environmental science. In this contribution we first derive the equations for the model error dynamics in the general case, and then for the particular situation of parametric error. We show how this deterministic description of the model error can be incorporated in sequential and variational data assimilation procedures. A numerical comparison with standard methods is given using low-order dynamical systems, prototypes of atmospheric circulation, and a realistic soil model. The deterministic approach proves to be very competitive with only minor additional computational cost. Most importantly, it offers a new way to address the problem of accounting for model error in data assimilation that can easily be implemented in systems of increasing complexity and in the context of modern ensemble-based procedures.

Keywords Data assimilation • Model error • Kalman filtering • Variational assimilation • Chaotic dynamics

1 Introduction

The prediction problem in geophysical fluid dynamics typically relies on two complementary elements: the model and the data. The mathematical model, and its discretized version, embodies our knowledge about the laws governing the system evolution, while the data are samples of the system's state. They give

---

A. Carrassi (✉)
NERSC – Nansen Environmental and Remote Sensing Center, Bergen, Norway
e-mail: alberto.carrassi@nersc.no

S. Vannitsem
RMI – Royal Meteorological Institute of Belgium, Brussels, Belgium
e-mail: svn@meteo.be

© Springer International Publishing Switzerland 2016   175
F. Ancona et al. (eds.), *Mathematical Paradigms of Climate Science*, Springer
INdAM Series 15, DOI 10.1007/978-3-319-39092-5_9

complementary information about the same object. The sequence of operations that merges model and data to obtain a possibly improved estimate of the flows state is usually known, in environmental science, as data assimilation [10, 23]. The physical and dynamical complexity of geophysical systems makes the data assimilation problem particularly involved.

The different information entering the data assimilation procedure, usually the model, the data and a background field representing the state estimate prior to the assimilation of new observations, are weighted according to their respective accuracy. Data assimilation in geophysics, particularly in numerical weather prediction (NWP) has experienced a long and fruitful stream of research in recent decades which has led to a number of advanced methods able to take full advantage of the increasing amount of available observations and to efficiently track and reduce the dynamical instabilities [15]. As a result the overall accuracy of the Earths system estimate and prediction, particularly the atmosphere, has improved dramatically.

Despite this trend of improvement, the treatment of model error in data assimilation procedures is still, in most instances, done following simple assumptions such as the absence of time correlation [21]. The lack of attention on model error is in part justified by the fact that on the time scale of NWP, where most of the geophysical data assimilation advancements have been originally concentrated, its influence is reasonably considered small as compared to the initial condition error that grows in view of the chaotic nature of the dynamics. Nevertheless, the improvement in data assimilation techniques and observational networks on the one hand, and the recent growth of interest in seasonal-to-decadal prediction on the other [14, 47], has placed model error, and its treatment in data assimilation, as a main concern and a key priority. A number of studies reflecting this concern have appeared, in the context of sequential and variational schemes [11, 25, 42, 43].

Two main obstacles toward the development of techniques taking into account model error sources are the huge size of the geophysical models and the wide range of possible model error sources. The former problem implies the need to estimate large error covariance matrices on the basis of the limited number of available observations. The second important issue is related to the multiple sources of modeling error, such as incorrect parametrisation, numerical discretization, and the lack of description of some relevant scale of motion. This latter problem has until recently limited the development of a general formulation for the model error dynamics. Model error is commonly *modeled* as an additive, stationary, zero-centered, Gaussian white noise process. This choice could be legitimate by the multitude of unknown error sources and the central limit theorem. However, despite this simplification, the size of geoscientific models still makes detailed estimation of the stochastic model error covariance impractical.

In the present contribution we describe an alternative approach in which the evolution of the model error is described based on a deterministic short-time approximation. The approximation is suitable for realistic applications and is used to estimate the model error contribution in the state estimate. The method is based on the theory of deterministic dynamics of the model error that was introduced recently by [33–35]. Using this approach it is possible to derive evolution equations for the moments of the model error statistics required in data assimilation procedures, that have been applied in the context of both sequential and variational data assimilation schemes, when model error originated from uncertain parameters or from unresolved scales.

We give here a review of the recent developments of the deterministic treatment of model error in data assimilation. To this end, we start by first formalizing the deterministic model error dynamics in Sect. 2. We show how general equations for the mean and covariance error can be obtained and discuss the parametric error as a special case. In Sects. 3 and 4 the incorporation of the short-time model error evolution laws is described in the context of the Extended Kalman filter and variational scheme respectively. These two types of assimilation procedures are significantly different and are summarized in the respective Sections along with the discussion on the consequences of the implementation of the model error treatment. We provide some numerical illustrations of the proposed approach, together with comparisons with other methods, for two prototypical low order chaotic systems widely used in theoretical studies in geosciences [28, 29] and a quasi-operational soil model [30].

New potential applications of the use of the deterministic model error treatment are currently under way and are summarized, along with a synopsis of the method, in the final discussion Sect. 5. These include soil data assimilation with the use of new observations and ensemble based procedures [15].

2   Formulation

Let the model at our disposal be represented as:

$$\frac{d\mathbf{x}(t)}{dt} = \mathbf{f}(\mathbf{x}, \lambda), \tag{1}$$

where $\mathbf{f}$ is typically a nonlinear function, defined in $\mathbf{R}^N$ and λ is a P-dimensional vector of parameters.

Model (1) is used to describe the evolution of a (unknown) true dynamics, *i.e. nature*, whose evolution is assumed to be given by the following coupled equations:

$$\frac{d\hat{\mathbf{x}}(t)}{dt} = \hat{\mathbf{f}}(\hat{\mathbf{x}}, \hat{\mathbf{y}}, \hat{\lambda}) \qquad \frac{d\hat{\mathbf{y}}(t)}{dt} = \hat{\mathbf{g}}(\hat{\mathbf{x}}, \hat{\mathbf{y}}, \hat{\lambda}) \tag{2}$$

where $\hat{\mathbf{x}}$ is a vector in $\mathbf{R}^N$, and $\hat{\mathbf{y}}$ is defined in $\mathbf{R}^L$ and may represent scales that are present in the real world, but are neglected in model (1); the unknown parameters $\hat{\lambda}$ have dimension P. The true state is thus a vector of dimension $N + L$. The model state vector $\mathbf{x}$ and the variable $\hat{\mathbf{x}}$ of the true dynamics span the same phase space although, given the difference in the functions $\mathbf{f}$ and $\hat{\mathbf{f}}$, they do not have the same attractor in general. The function $\mathbf{f}$ can have an explicit dependence on time but it is dropped here to simplify the notation.

When using model (1) to describe the evolution of $\hat{\mathbf{x}}$, estimation error can arise from the uncertainty in the initial conditions at the resolved scale ($\mathbf{x}(t_0) \neq \hat{\mathbf{x}}(t_0)$) and from the approximate description of the nature afforded by (1) which is referred as model error.

2.1 General Description of Model Error Dynamics

Following the approach outlined in [34], we derive the evolution equations of the dominant moments, mean and covariance, of the estimation error $\delta\mathbf{x} = \mathbf{x} - \hat{\mathbf{x}}$ in the resolved scale (i.e. in $\mathbf{R}^N$). The formal solutions of (1) and (2) read respectively:

$$\mathbf{x}(t) = \mathbf{x}_0 + \int_0^t d\tau \mathbf{f}(\mathbf{x}(\tau), \boldsymbol{\lambda}) \tag{3}$$

$$\hat{\mathbf{x}}(t) = \hat{\mathbf{x}}_0 + \int_0^t d\tau \hat{\mathbf{f}}(\hat{\mathbf{x}}(\tau), \hat{\mathbf{y}}(\tau), \hat{\boldsymbol{\lambda}}) \tag{4}$$

where $\mathbf{x}_0 = \mathbf{x}(t_0)$, and $\hat{\mathbf{x}}_0 = \hat{\mathbf{x}}(t_0)$. By taking the difference between (3) and (4), and averaging over an ensemble of perturbations around a reference state, we get the formal solution for the mean error, the bias:

$$< \delta\mathbf{x}(t) > = < \delta\mathbf{x}_0 > + \int_0^t d\tau < \mathbf{f}(\mathbf{x}(\tau), \boldsymbol{\lambda}) - \hat{\mathbf{f}}(\hat{\mathbf{x}}(\tau), \hat{\mathbf{y}}(\tau), \hat{\boldsymbol{\lambda}}) > \tag{5}$$

with $\delta\mathbf{x}_0 = \mathbf{x}_0 - \hat{\mathbf{x}}_0$. Two types of averaging could be performed, one over a set of initial conditions sampled on the attractor of the system, and/or a set of perturbations around one specific initial state selected on the system's attractor. In data assimilation, the second is more relevant since one is interested in the local evaluation of the uncertainty. However, in many situations the first one is used to get statistical information on covariances quantities, as will be illustrated in this Chapter. For clarity, we will refer to $< . >$ as the local averaging, and to $<< . >>$ for an averaging over a set of initial conditions sampled over the attractor of the system. In this section, we will only use $< . >$ for clarity, but it also extends to the other averaging. We will use the other notation $<< . >>$ when necessary.

In the hypothesis that the initial condition is unbiased, $< \delta\mathbf{x}_0 > = 0$, Eq. (5) gives the evolution equation of the bias due to the model error, usually refers to as drift in climate prediction context. The important factor driving the drift is the difference between the true and modeled tendency fields, $< \mathbf{f}(\mathbf{x}(\tau), \boldsymbol{\lambda}) - \hat{\mathbf{f}}(\hat{\mathbf{x}}(\tau), \hat{\mathbf{y}}(\tau), \hat{\boldsymbol{\lambda}}) >$. Expanding (5) in Taylor series around $t_0 = 0$ up to the first non-trivial order, and using unbiased initial conditions, it reads:

$$\mathbf{b}^m(t) = < \delta\mathbf{x}(t) > \approx < \mathbf{f}(\mathbf{x}(\tau), \boldsymbol{\lambda}) - \hat{\mathbf{f}}(\hat{\mathbf{x}}(\tau), \hat{\mathbf{y}}(\tau), \hat{\boldsymbol{\lambda}}) > t . \tag{6}$$

Equation (6) gives the evolution of the bias, $\mathbf{b}^m$ (the drift) in the short-time approximation and the subscript m stands for model error-related bias. It is important to remark that in the case of classical stochastic model error treatment, and in the hypothesis of unbiased initial condition error, $\mathbf{b}^m = 0$.

Similarly, by taking the expectation of the external product of the error anomalies $\delta\mathbf{x}$ by themselves, we have:

$$\mathbf{P}(t) = <\{\delta\mathbf{x}(t)\}\{\delta\mathbf{x}(t)\}^T> = <\{\delta\mathbf{x}_0\}\{\delta\mathbf{x}_0\}^T> +$$

$$<\{\delta\mathbf{x}_0\}\{\int_0^t d\tau[\mathbf{f}(\mathbf{x}(\tau),\lambda) - \hat{\mathbf{f}}(\hat{\mathbf{x}}(\tau),\hat{\mathbf{y}}(\tau),\hat{\lambda})]\}^T> +$$

$$<\{\int_0^t d\tau[\mathbf{f}(\mathbf{x}(\tau),\lambda) - \hat{\mathbf{f}}(\hat{\mathbf{x}}(\tau),\hat{\mathbf{y}}(\tau),\hat{\lambda})]\}\{\delta\mathbf{x}_0\}^T> +$$

$$\int_0^t d\tau \int_0^t d\tau' <\{\mathbf{f}(\mathbf{x}(\tau),\lambda) - \hat{\mathbf{f}}(\hat{\mathbf{x}}(\tau),\hat{\mathbf{y}}(\tau),\hat{\lambda})\}\{\mathbf{f}(\mathbf{x}(\tau'),\lambda) - \hat{\mathbf{f}}(\hat{\mathbf{x}}(\tau'),\hat{\mathbf{y}}(\tau'),\hat{\lambda})\}^T> .$$
$$(7)$$

Equation (7) describes the time evolution of the estimation error covariance in the resolved scale. The first term, that does not depend on time, represents the covariance of the initial error. The two following terms account for the correlation between the error in the initial condition and the model error, while the last term combines the effect of both errors on the evolution of the estimation error covariance.

Let us focus on the last term of Eq. (7) denoted as,

$$\mathbf{P}(t) = \int_0^t d\tau \int_0^t d\tau' <\{\mathbf{f}(\mathbf{x}(\tau),\lambda) - \hat{\mathbf{f}}(\hat{\mathbf{x}}(\tau),\hat{\mathbf{y}}(\tau),\hat{\lambda})\}\{\mathbf{f}(\mathbf{x}(\tau'),\lambda) - \hat{\mathbf{f}}(\hat{\mathbf{x}}(\tau'),\hat{\mathbf{y}}(\tau'),\hat{\lambda})\}^T>$$
$$(8)$$

The amplitude and structure of this covariance depends on the dynamical properties of the difference of the nature and model tendency fields. Assuming that these differences are correlated in time, we can expand (8) in a time series up to the first nontrivial order around the arbitrary initial time $t_0 = 0$, and gets:

$$\mathbf{P}^m(t) \approx <\{\mathbf{f}(\mathbf{x}_0,\lambda) - \hat{\mathbf{f}}(\hat{\mathbf{x}}_0,\hat{\mathbf{y}}_0,\hat{\lambda})\}\{\mathbf{f}(\mathbf{x}_0,\lambda) - \hat{\mathbf{f}}(\hat{\mathbf{x}}_0,\hat{\mathbf{y}}_0,\hat{\lambda})\}^T> t^2 = \mathbf{Q}t^2 \quad (9)$$

where $\mathbf{Q}$ is the model error covariance matrix at initial time. Note again that, if the terms $\mathbf{f} - \hat{\mathbf{f}}$ are represented as white-noise process, the short-time evolution of $\mathbf{P}(t)$ is bound to be linear instead of quadratic. This distinctive feature is relevant in data assimilation applications where model error is often assumed to be uncorrelated in time, a choice allowing for a reduction of the computational cost associated with certain types of algorithms [5, 42].

2.2 Model Error Due to Parameter Uncertainties

We assume for simplicity that the model resolves all scales present in the reference system. Under the aforementioned hypothesis that the model and the true trajectories span the same phase space, nature dynamics, (2), can be rewritten as:

$$\frac{d\hat{\mathbf{x}}(t)}{dt} = \mathbf{f}(\hat{\mathbf{x}}, \hat{\lambda}) + \varepsilon\mathbf{h}(\hat{\mathbf{x}}, \gamma) \,. \tag{10}$$

The function $\mathbf{h}$, which has the same order of magnitude of $\mathbf{f}$ and is scaled by the dimensionless parameter ε, accounts for all other extra terms not included in the model and depends on the resolved variable $\hat{\mathbf{x}}$ and on a set of additional parameters γ. In a more formal description, this $\mathbf{h}$ would correspond to a function relating the variables $\hat{\mathbf{x}}$ and $\hat{\mathbf{y}}$ under an adiabatic elimination [34]. We are interested here in a situation in which the main component of the nature dynamics is well captured by the model so that $\varepsilon \ll 1$, and the extra terms described by $\mathbf{h}$ are neglected. We concentrate in a situation in which model error is due only to uncertainties in the specification of the parameters appearing in the evolution law $\mathbf{f}$. This formulation accounts, for instance, for errors in the description of some physical processes (dissipation, external forcing, etc.) represented by the parameters.

An equation for the evolution of the state estimation error $\delta\mathbf{x}$ can be obtained by taking the difference between the first rhs term in (10) and (1). The evolution of $\delta\mathbf{x}$ depends on the error estimate at the initial time $t = t_0$ (initial condition error $\delta\mathbf{x}(t_0) = \delta\mathbf{x}_0$) and on the model error. If $\delta\mathbf{x}$ is "small", the linearized dynamics provides a reliable approximation of the actual error evolution. The linearization is made along a model trajectory, solution of (1), by expanding, to first order in $\delta\mathbf{x}$ and $\delta\lambda = \lambda - \hat{\lambda}$, the difference between Eqs. (10) and (1):

$$\frac{d\delta\mathbf{x}}{dt} = \frac{\partial\mathbf{f}}{\partial\mathbf{x}}|_{\mathbf{x}}\delta\mathbf{x} + \frac{\partial\mathbf{f}}{\partial\lambda}|_{\lambda}\delta\lambda \,. \tag{11}$$

The first partial derivative on the rhs of (11) is the Jacobian of the model dynamics evaluated along its trajectory. The second term, which corresponds to the model error, will be denoted $\delta\mu$ hereafter to simplify the notation; $\delta\mu = \frac{\partial\mathbf{f}}{\partial\lambda}|_{\lambda}\delta\lambda$

The solution of (11), with initial condition $\delta\mathbf{x}_0$ at $t = t_0$, reads:

$$\delta\mathbf{x}(t) = \mathbf{M}_{t,t_0}\delta\mathbf{x}_0 + \int_{t_0}^{t} d\tau\mathbf{M}_{t,\tau}\delta\mu(\tau)$$

$$= \delta\mathbf{x}^{ic}(t) + \delta\mathbf{x}^{m}(t) \tag{12}$$

with $\mathbf{M}_{t,t_0}$ being the fundamental matrix (the propagator) relative to the linearized dynamics along the trajectory between t_0 and t. We point out that $\delta\mu$ and $\mathbf{M}_{t,\tau}$ in (12) depend on τ (the integration variable) through the state variable $\mathbf{x}$. Equation (12) states that, in the linear approximation, the error in the state estimate is given by the sum of two terms, the evolution of initial condition error, $\delta \mathbf{x}^{ic}$, and the model error, $\delta \mathbf{x}^m$. The presence of the fundamental matrix $\mathbf{M}$ in the expression for $\delta \mathbf{x}^m$ suggests that the instabilities of the flow plays a role in the dynamics of model error.

Let us now apply the expectation operator to (12) defined locally around the reference trajectory, by sampling over an ensemble of initial conditions and model errors, and the equation for the mean estimation error along a reference trajectory reads:

$$< \delta \mathbf{x}(t) >= \mathbf{M}_{t,t_0} < \delta \mathbf{x}_0 > + \int_{t_0}^{t} d\tau \mathbf{M}_{t,\tau} < \delta \boldsymbol{\mu}(\tau) >$$

$$=< \delta \mathbf{x}^{ic} > + < \delta \mathbf{x}^m > . \tag{13}$$

In a perfect model scenario an unbiased state estimate at time t_0 ($< \delta \mathbf{x}_0 >= 0$) will evolve, under the linearized dynamics, into an unbiased estimate at time t. In the presence of model error and, depending on its properties, an initially unbiased estimate can evolve into a biased one with $< \delta \boldsymbol{\mu}(t) >$ being the key factor.

The dynamics of the state estimation error covariance matrix can be obtained by taking the expectation of the outer product of $\delta \mathbf{x}(t)$ with itself. Assuming that the estimation error bias is known and removed from the background error, we get:

$$\mathbf{P}(t) =< \delta \mathbf{x}(t) \delta \mathbf{x}(t)^T >$$

$$= \mathbf{P}^{ic}(t) + \mathbf{P}^m(t) + \mathbf{P}^{corr}(t) + (\mathbf{P}^{corr})^T(t) \tag{14}$$

where:

$$\mathbf{P}^{ic}(t) = \mathbf{M}_{t,t_0} < \delta \mathbf{x}_0 \delta \mathbf{x}_0{}^T > \mathbf{M}_{t,t_0}^T \tag{15}$$

$$\mathbf{P}^m(t) = \int_{t_0}^{t} d\tau \int_{t_0}^{t} d\tau' \mathbf{M}_{t,\tau} < \delta \boldsymbol{\mu}(\tau) \delta \boldsymbol{\mu}(\tau')^T) > \mathbf{M}_{t,\tau'}^T \tag{16}$$

$$\mathbf{P}^{corr}(t) = \mathbf{M}_{t,t_0} < (\delta \mathbf{x}_0) \left(\int_{t_0}^{t} d\tau \mathbf{M}_{t,\tau} \delta \boldsymbol{\mu}(\tau) \right)^T > . \tag{17}$$

The four terms on the r.h.s. of (14) give the contribution to the estimation error covariance at time t due to the initial condition, model error and their cross correlation, respectively. These integral equations are of little practical use for any realistic nonlinear systems, let alone the big models used in environmental prediction. A suitable expression can be obtained by considering their short-time approximations through a Taylor expansion around t_0. We proceed by expanding (12) in Taylor series, up to the first non trivial order, only for the model error term $\delta \mathbf{x}^m$ while keeping the initial condition term, $\delta \mathbf{x}^{ic}$, unchanged. In this case, the model error $\delta \mathbf{x}^m$

evolves linearly with time according to:

$$\delta \mathbf{x}^m \approx \delta \boldsymbol{\mu}_0 (t - t_0) \tag{18}$$

where $\delta \boldsymbol{\mu}(t_0) = \delta \boldsymbol{\mu}_0$.

By adding the initial condition error term, $\delta \mathbf{x}^{ic}$, we get a short time approximation of (12):

$$\delta \mathbf{x}(t) \approx \mathbf{M}_{t,t_0} \delta \mathbf{x}_0 + \delta \boldsymbol{\mu}_0 (t - t_0) . \tag{19}$$

For the mean error we get:

$$\mathbf{b}^m(t) \approx <\delta \mathbf{x}(t)> \approx \mathbf{M}_{t,t_0} <\delta \mathbf{x}_0> + <\delta \boldsymbol{\mu}_0> (t - t_0) . \tag{20}$$

Therefore, as long as $<\delta \boldsymbol{\mu}_0>$ is different from zero, the bias due to parametric error evolves linearly for short-time, otherwise the evolution is conditioned by higher orders of the Taylor expansion. Note that the two terms in the short time error evolution (19) and (20), are not on equal footing since, in contrast to the model error term, which been expanded up to the first nontrivial order in time, the initial condition error evolution contains all the orders of times $(t, t^2, \ldots, t^n)$. The point is that, as explained below, we intend to use these equations to model the error evolution in conjunction with the technique of data assimilation for which the full matrix $\mathbf{M}$, or an amended ensemble based approximation, is already available.

Taking the expectation value of the external product of (19) by itself and averaging, we get:

$$\mathbf{P}(t) \approx \mathbf{M}_{t,t_0} <\delta \mathbf{x}_0 \delta \mathbf{x}_0^T> \mathbf{M}_{t,t_0}^T +$$

$$+ [<\delta \boldsymbol{\mu}_0 \delta \mathbf{x}_0^T> \mathbf{M}_{t,t_0}^T + \mathbf{M}_{t,t_0} <\delta \mathbf{x}_0 \delta \boldsymbol{\mu}_0^T>](t - t_0) + <\delta \boldsymbol{\mu}_0 \delta \boldsymbol{\mu}_0^T> (t - t_0)^2 . \tag{21}$$

Equation (21) is the short time evolution equation, in this linearized setting, for the error covariance matrix in the presence of both initial condition and parametric model errors.

3 Deterministic Model Error Treatment in the Extended Kalman Filter

We describe here two formulations of the extended Kalman filter (EKF) incorporating a model error treatment. The Short-Time-Extended-Kalman-Filter, ST-EKF [8] accounts for model error through an estimate of its contribution to the assumed forecast error statistics. In the second formulation, the Short-Time-Augmented-Extended-Kalman-Filter, ST-AEKF [6], the state estimation in the EKF is accompanied with the estimation of the uncertain parameters. This is done in the context of a general framework known as state augmentation [21]. In both cases model error is treated as a deterministic process implying that the dynamical laws described in the previous section are incorporated, at different stages, in the filter formulations.

The EKF extends, to nonlinear dynamics, the classical Kalman filter (KF) for linear dynamics [22]. The algorithm is sequential in the sense that a prediction of the system's state is updated at discrete times, when observations are present. The state update, the analysis, is then taken as the initial condition for the subsequent prediction up to the next observation time. The EKF, as well as the standard KF for linear dynamics, is derived in the hypothesis of Gaussian errors whose distributions can thus be fully described using only the first two moments, the mean and the covariance. Although this can represent a very crude approximation, especially for nonlinear systems, it allows for a dramatic reduction of the cost and difficulties involved in the time propagation of the full error distribution.

The model equations can conveniently be written in terms of a discrete mapping from time t_k to t_{k+1}:

$$\mathbf{x}^f_{k+1} = \mathcal{M} \mathbf{x}^a_k \tag{22}$$

where $\mathbf{x}^f$ and $\mathbf{x}^a$ are the forecast and analysis states respectively and $\mathcal{M}$ is the nonlinear model forward operator (the resolvent of (1)).

Let us assume that a set of M noisy observations of the true system (2), stored as the components of an M-dimensional observation vector $\mathbf{y}^o$, is available at the regularly spaced discrete times $t_k = t_0 + k\tau$, $k = 1, 2 \ldots$, with τ being the assimilation interval, so that:

$$\mathbf{y}^o_k = \mathcal{H}(\hat{\mathbf{x}}_k) + \boldsymbol{\varepsilon}^o_k \tag{23}$$

where ε^0_k is the observation error at time t_k, assumed here to be Gaussian with known covariance matrix $\mathbf{R}$ and uncorrelated in time. $\mathcal{H}$ is the (possibly nonlinear) observation operator which maps from model to observation space (i.e. from model to observed variables) and may involve spatial interpolations as well as transformations based on physical laws for indirect measurements [20].

For the EKF, as well as for most least-square based assimilation schemes, the analysis state update equation at an arbitrary analysis time t_k, reads [21]:

$$\mathbf{x}^a = [\mathbf{I} - \mathbf{KH}]\,\mathbf{x}^f + \mathbf{K}\mathbf{y}^o \tag{24}$$

where the time indexes are dropped to simplify the notation. The analysis error covariance, $\mathbf{P}^a$, is updated through:

$$\mathbf{P}^a = [\mathbf{I} - \mathbf{KH}]\,\mathbf{P}^f . \tag{25}$$

The $I \times M$ *gain* matrix $\mathbf{K}$ is given by:

$$\mathbf{K} = \mathbf{P}^f \mathbf{H}^T \left[\mathbf{H} \mathbf{P}^f \mathbf{H}^T + \mathbf{R} \right]^{-1} \tag{26}$$

where $\mathbf{P}^f$ is the $I \times I$ forecast error covariance matrix and $\mathbf{H}$ the linearized observation operator (a $M \times I$ real matrix). The analysis update is thus based on two complementary sources of information, the observations, $\mathbf{y}^o$, and the forecast $\mathbf{x}^f$. The errors associated to each of them are assumed to be uncorrelated and fully described by the covariance matrices $\mathbf{R}$ and $\mathbf{P}^f$, respectively.

In the EKF, the forecast error covariance matrix, $\mathbf{P}^f$, is obtained by linearizing the model around its trajectory between two successive analysis times t_k and $t_{k+1} = t_k + \tau$. In the standard formulation of the EKF model error is assumed to be a random uncorrelated noise whose effect is modeled by adding a model error covariance matrix, $\mathbf{P}^m$, at the forecast step so that [21]:

$$\mathbf{P}^f = \mathbf{M} \mathbf{P}^a \mathbf{M}^T + \mathbf{P}^m . \tag{27}$$

In practice the matrix $\mathbf{P}^m$ should be considered as a measure of the variability of the noise sequence. This approach has been particularly attractive in the past in view of its simplicity and because of the lack of more refined model for the model error dynamics. Note that while $\mathbf{P}^f$ is propagated in time and is therefore flow dependent, $\mathbf{P}^m$ is defined once for all and it is then kept constant.

3.1 Short Time Extended Kalman Filter: ST-EKF

We study here the possibility of estimating the model error covariance, $\mathbf{P}^m$, on a deterministic basis [8]. The approach uses the formalism on model error dynamics outlined in Sect. 2.

Model error is regarded as a time-correlated process and the short-time evolution laws (6) and (9) are used to estimate the bias, $\mathbf{b}^m$, and the model error covariance matrix, $\mathbf{P}^m$, respectively. The adoption of the short-time approximation is also legitimated by the sequential nature of the EKF, and an important practical concern is the ratio between the duration of the short-time regime and the length of the assimilation interval τ over which the approximation is used [34].

A key issue is the estimation of the two first statistical moments of the tendency mismatch, $\mathbf{f} - \hat{\mathbf{f}}$, required in (6) and in (9), respectively. The problem is addressed assuming that a reanalysis dataset of relevant geophysical fields is available and is used as a proxy of the nature evolution. Reanalysis programs constitute the best-possible estimate of the Earth system over an extended period of time, obtained using an homogeneous model and data assimilation procedure, and are of paramount importance in climate diagnosis (see e.g. [12]).

Let us suppose to have access to such a reanalysis which includes the analysis, $\mathbf{x}_r^a$, and the forecast field, $\mathbf{x}_r^f$, so that $\mathbf{x}_r^f(t_j + \tau_r) = \mathcal{M}\mathbf{x}_r^a(t_j)$, and τ_r is the assimilation interval of the data assimilation scheme used to produce the reanalysis; the suffix r stands for reanalysis. Under this assumption the following approximation is made:

$$\mathbf{f}(\mathbf{x}, \lambda) - \hat{\mathbf{f}}(\hat{\mathbf{x}}, \hat{\mathbf{y}}, \lambda, \hat{\mathbf{e}}) = \frac{d\mathbf{x}}{dt} - \frac{d\hat{\mathbf{x}}}{dt} \approx$$

$$\frac{\mathbf{x}_r^f(t + \tau_r) - \mathbf{x}_r^a(t)}{\tau_r} - \frac{\mathbf{x}_r^a(t + \tau_r) - \mathbf{x}_r^a(t)}{\tau_r} = \frac{\mathbf{x}_r^f(t + \tau_r) - \mathbf{x}_r^a(t + \tau_r)}{\tau_r} = -\frac{\delta\mathbf{x}_r^a}{\tau_r}.$$

$$(28)$$

The difference between the analysis and the forecast, $\delta\mathbf{x}_r^a$, is usually referred, in data assimilation literature, to as the *analysis increment*. From (28) we see that the vector of analysis increments can be used to estimate the difference between the model and the true tendencies. A similar approach was originally introduced by Leith [26], and it has been used recently to account for model error in data assimilation [19].

Note that the estimate (28) neglects the analysis error, so that its accuracy is connected to that of the data assimilation algorithm used to produce the reanalysis, which is in turn related to the characteristics of the observational network such as number, distribution and frequency of the observations. However this error is present and acts as an initial condition error, a contribution which is already accounted for in the EKF update by the forecast error covariance, $\mathbf{P}^f$. As a consequence when (9) is used to estimate only the model error component, an overestimation is expected that can be overcome by an optimal tuning of the amplitude of $\mathbf{b}^m$ and $\mathbf{P}^m$.

The most straightforward way to estimate the bias due to model error using (28) in (6), so that at analysis time it reads:

$$\mathbf{b}^m = -\sqrt{\alpha} < \delta\mathbf{x}_r^a > \frac{\tau}{\tau_r}. \tag{29}$$

The bias is then removed from the forecast field before the latter enters the EKF analysis update, (24). The scalar term α is a tunable coefficient aimed at optimizing the bias size to account for the expected overestimation connected with the use of (28). In a similar way the model error contribution to the forecast error covariance can be estimated taking the external product of (28) after removing the mean and reads:

$$\mathbf{P}^m = \alpha < \{\delta\mathbf{x}_r^a - < \delta\mathbf{x}_r^a >\}\{\delta\mathbf{x}_r^a - < \delta\mathbf{x}_r^a >\}^T > \frac{\tau^2}{\tau_r^2}. \tag{30}$$

We consider now the particular case of parametric error. The forecast error covariance $\mathbf{P}^f$, is estimated using the short-time evolution (21) where the correlation terms are neglected and the model error covariance, $\mathbf{P}^m$ is evolved quadratically in the intervals between observations. An additional advantage is that $\mathbf{P}^m$ can be straightforwardly adapted to different assimilation intervals and for the assimilation of asynchronous observations. At analysis times the forecast error bias due to the model error, $\mathbf{b}^m$, can be estimated on the basis of the short-time approximation (20):

$$\mathbf{b}^m =< \delta\mathbf{x}^m >\approx< \delta\mu_o > \tau . \tag{31}$$

By neglecting the correlation terms and dropping the time dependence for convenience, Eq. (21) can be rewritten as:

$$\mathbf{P}^f = \mathbf{M}\mathbf{P}^a\mathbf{M}^T + < \delta\mu_o\delta\mu_o^T > \tau^2 = \mathbf{M}\mathbf{P}^a\mathbf{M}^T + \mathbf{Q}\tau^2 = \mathbf{M}\mathbf{P}^a\mathbf{M}^T + \mathbf{P}^m \tag{32}$$

where $\mathbf{P}^a$ is the analysis error covariance matrix, as estimated at the last analysis time, and

$$\mathbf{P}^m = \mathbf{Q}\tau^2 =< \delta\mu_o\delta\mu_o^T > \tau^2. \tag{33}$$

An essential ingredient of the ST-EKF in the case of parametric error is the matrix $\mathbf{Q}$: it embeds the information on the model error through the unknown parametric error $\delta\lambda$ and the parametric functional dependence of the dynamics. In [8] it was supposed that some a-priori information on the model error was at disposal and could be used to prescribe $< \delta\mu_o >$ and $\mathbf{Q}$ then used to compute $\mathbf{b}^m$ and $\mathbf{P}^m$ required by the ST-EKF. The availability of information on the model error, which may come in practice from the experience of modelers, is simulated by estimating $< \delta\mu_o >$ and $\mathbf{Q}$ averaging over a large sample of states on the system's attractor as,

$$\mathbf{b}^m = << \delta\mu_o >> \tag{34}$$

$$\mathbf{Q} = << \delta\mu_o\delta\mu_o^T >> \tag{35}$$

The same assumption is adopted here in the numerical applications with the ST-EKF described in Sect. 3.2.1.

In summary, in the ST-EKF, either in general or in the parameteric error case, once $\mathbf{b}^m$ and $\mathbf{P}^m$ are estimated (with (29)–(30) or (31), (32) and (33), respectively) they are then kept constant along the entire assimilation cycle. Model error is thus repeatedly corrected in the subspace spanned by the range of $\mathbf{P}^m$ where it is supposed to be confined. This choice reflects the assumption that the impact of model uncertainty on the forecast error does not fluctuate too much along the analysis cycle. Finally, in the ST-EKF, the forecast field and error covariance are transformed according to:

$$\mathbf{x}^f \Longrightarrow \mathbf{x}^f - \mathbf{b}^m, \tag{36}$$

$$\mathbf{P}^f \Longrightarrow \mathbf{P}^f + \mathbf{P}^m . \tag{37}$$

These new first guess and forecast error covariance, (36) and (37), are then used in the EKF analysis formulas (24)–(25).

3.1.1 Numerical Results with ST-EKF: Error Due to Unresolved Scale

We show here numerical results of the ST-EKF for the case of model error arising from the lack of description of a scale of motion. The case of parametric error is numerically tested in Sect. 3.2.1. A standard approach, known in geosciences as observation system simulation experiments (OSSE), is adopted here [2]. This experimental setup is based on a twin model configuration in which a trajectory, solution of the system taken to represent the actual dynamics, is sampled to produce synthetic observations. A second model provides the forecasting system that assimilates the observations.

As a prototype of two-scales chaotic dynamics we consider the model introduced by [29], whose equations read:

$$\frac{dx_i}{dt} = (x_{i+1} - x_{i-2})x_{i-1} - x_i + F - \frac{hc}{b}\sum_{j=1}^{10} y_{j,i}, \qquad i = \{1, \ldots, 36\} \tag{38}$$

$$\frac{dy_{j,i}}{dt} = -cby_{j+1,i}(y_{j+2,i} - y_{j-1,i}) - cy_{j,i} + \frac{hc}{b}x_i, \qquad j = \{1, \ldots, 10\}. \tag{39}$$

The model possesses two distinct scales of motion evolving according to (38) and (39), respectively. The large/slow scale variable, x_i, represents a generic meteorological variable over a circle at fixed latitude. In both set of equations, the quadratic term simulates the advection, the second rhs term the internal dissipation, while the constant term in (38) plays the role of the external forcing. The two scales are coupled in such a way that the small/fast scale variables $y_{j,i}$ inhibit the larger ones, while the opposite occurs for the effect of the variables x_i on $y_{j,i}$. According to [29] the variables $y_{j,i}$ can be taken to represent some convective-scale quantity, while the variables x_i favor this convective activity. The model parameters are set as in [29]: $c = b = 10$, which makes the variables x_i to vary ten times slower than $y_{j,i}$, with amplitudes ten times larger, while $F = 10$ and $h = 1$. With this choice, the dynamics is chaotic. The numerical integration have been performed using a fourth-order Runge-Kutta scheme with a time step of 0.0083 units, corresponding to 1 h of simulated time.

In the experiments the full Eqs. (38)–(39), are taken to represent the truth, while the model sees only the slow scale and its equations are given by (38) without the last term. A network of $M = 12$ regularly spaced noisy observations of $\mathbf{x}$ is simulated by sampling the reference true trajectory and adding a Gaussian random observation error. We first generate a long record of analysis for the state vector, $\mathbf{x}$, which constitutes the reanalysis dataset. The EKF algorithm is run for 10 years with assimilation interval $\tau_r = 6$ h, and observation variance set to 5 % of the system's climate variance. From this long integration we extract the record of analysis increments required in (29) and (30).

An illustration of the impact of the proposed treatment of the model error is given in Fig. 1, which shows a 30 days long assimilation cycle. The upper panel displays the true large scale variable x_{16} (blue line), the corresponding estimates obtained

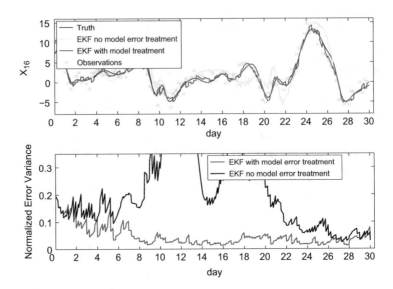

Fig. 1 *Top Panel*: Model variable $\hat{x}_{16}$ for the truth (*blue*), EKF without model error treatment $\mathbf{Q} = 0$, (*yellow*), EKF with model error treatment (*red*) and observations (*green*). *Bottom Panel*: Estimation error variance, normalized with respect to the system's climate variance, as a function of time (Republished with permission of World Scientific Publishing Co., from [7]; permission conveyed through Copyright Clearance Center, Inc)

with the ST-EKF and the EKF without the model error treatment (red and yellow lines respectively) and the observations (green marks). In the experiments without model error treatment, the model error covariance matrix $\mathbf{Q}$ is set to 0. The error variance of the EKF estimates are shown in the bottom panel. From the top panel we see the improvement in the tracking of the true trajectory obtained by implementing the proposed model error treatment; this is particularly evident in the proximity of the maxima and minima of the true signal. The benefit is further evident by looking at the estimated error variance which undergoes a rapid convergence to values close or below the observation error.

A common practical procedure used to account for model error in KF-like and ensemble-based schemes, is the multiplicative covariance inflation [1]. The forecast error covariance matrix $\mathbf{P}^f$ is multiplied by a scalar factor and thus inflated while keeping its spatial structure unchanged, so that $\mathbf{P}^f \rightarrow (1 + \rho)\mathbf{P}^f$ before its use in the analysis update, (24). We have followed the same procedure here and have optimized the EKF by tuning the inflation factor ρ; the results are reported in Fig. 2a which shows the normalized estimation error variance as a function of ρ. The experiments last for 210 days, and the results are averaged in time, after an initial transient of 30 days, and over a sample of 100 random initial conditions. The best performance is obtained by inflating $\mathbf{P}^f$ by 9 % of its original amplitude and the estimation error variance is about 6 % of the system's climate variance, slightly

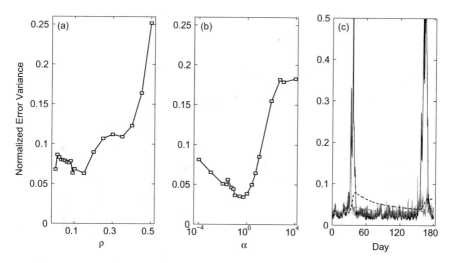

Fig. 2 Averaged normalized estimation error variance as a function of (**a**) the inflation factor ρ, (**b**) the coefficient α (log scale in the x-axis), and (**c**) time evolution of the normalized estimation error variance for the case $\rho = 0.09$ (*black*) and $\alpha = 0.5$ (*red*) (the time running mean is displayed with *dashed lines*) (Republished with permission of World Scientific Publishing Co., from [7]; permission conveyed through Copyright Clearance Center, Inc)

above the observation error variance. Note that when $\rho = 0$ filter divergence occurs in some of the 100 experiments.

We now test the sensitivity of the ST-EKF to the multiplicative coefficient α in (29) and (30). The results are reported in Fig. 2b, which shows the estimation error variance as a function of α. As above the averages are taken over 180 days and over the same ensemble of 100 random initial conditions. The important feature is the existence of a range of values of α, for which the estimation error is below the observation error level. Note that for $\alpha = 1$, the estimation error is about 4 % of the climate's variance, below the observational accuracy. This result highlights the accuracy of the estimate of $\mathbf{P}^m$ despite the simplifying assumptions such as the one associated with the correlation between model error and initial condition error and the use of the reanalysis field as a proxy of the actual true trajectory. Interestingly, the best performance is obtained with $\alpha = 0.5$, in agreement with the expected overestimation connected with the use of (28).

In Fig. 2c we explicitly compare the EKF with the optimal inflation for $\mathbf{P}^f$, ($\rho = 0.09$, $\mathbf{P}^m = \rho \mathbf{P}^f$), with the EKF implementing the model error treatment through the matrix $\mathbf{P}^m$ estimated according to (30) and tuned with the optimal values of the scalar coefficient $\alpha = 0.5$. The figure displays the estimation error variance as a function of time. Note in particular the ability of the ST-EKF filter, using $\mathbf{P}^m$, to keep estimation error small even in correspondence with the two large deviations experienced by the EKF employing the forecast error covariance inflation.

3.2 Short Time Augmented Extended Kalman Filter: ST-AEKF

Let us now turn to the state-augmentation approach. In this case we will assume that model errors arise from mis-specifications of some parameters, so that the theory depicted in Sect. 2.2 can be used. This view restricts us to parametric errors, but it also reflects our limited knowledge of the sub-grid scale processes that are only represented through parametrisation schemes for which only a set of parameters is accessible.

A straightforward theory exists for the estimation of the uncertain parameters along with the system's state. The approach, commonly known as state-augmentation [21], consists in defining an augmented dynamical system which allocates, along with the system's state, the model parameters to be estimated. The analysis update is then applied to this new augmented dynamical system. Our aim here is to use the state-augmentation approach in conjunction with the deterministic formulation of the model error dynamics.

The dynamical system (22), the forecast model, is augmented with the model parameters, as follows:

$$\mathbf{z}^f = \begin{bmatrix} \mathbf{x}^f \\ \boldsymbol{\lambda}^f \end{bmatrix} = \mathscr{F}\mathbf{z}^a = \begin{bmatrix} \mathscr{M}\mathbf{x}^a \\ \mathscr{F}^\lambda \boldsymbol{\lambda}^a \end{bmatrix} \tag{40}$$

where $\mathbf{z} = (\mathbf{x}, \boldsymbol{\lambda})$ is the augmented state vector. The augmented dynamical system $\mathscr{F}$ includes the dynamical model for the system's state, $\mathscr{M}$, and a dynamical model for the parameters $\mathscr{F}^\lambda$. In the absence of additional information, a persistence model for $\mathscr{F}^\lambda$ is usually assumed so that $\mathscr{F}^\lambda = \mathbf{I}$ and $\boldsymbol{\lambda}^f_{t_{k+1}} = \boldsymbol{\lambda}^a_{t_k}$. Recently, a temporally smoothed version of the persistence model has been used in the framework of a square root filter [46]. The state-augmented formulation is also successfully applied in the context of the general class of ensemble-based data assimilation procedures [39].

By proceeding formally as for Eq. (12) we can write the linearized error evolution for the augmented system, in an arbitrary assimilation interval, $\tau = t_{k+1} - t_k$, with initial condition given by the augmented state analysis error, $\delta\mathbf{z}^a = (\delta\mathbf{x}^a, \delta\boldsymbol{\lambda}^a) = (\mathbf{x}^a - \hat{\mathbf{y}}, \boldsymbol{\lambda}^a - \hat{\boldsymbol{\lambda}})$:

$$\delta\mathbf{z}^f \approx (\delta\mathbf{x}^f, \delta\boldsymbol{\lambda}^f) = (\mathbf{M}\delta\mathbf{x}^a + \int_t^{t+\tau} ds \mathbf{M}_{t,s} \delta\boldsymbol{\mu}^a(s), \delta\boldsymbol{\lambda}^a) \tag{41}$$

with $\delta\boldsymbol{\mu}^a = (\frac{\partial \mathbf{f}}{\partial \lambda}|_{\lambda^a})\delta\boldsymbol{\lambda}^a$. The parametric error $\delta\boldsymbol{\lambda}^f_{t_{k+1}} = \delta\boldsymbol{\lambda}^a_{t_k}$ is constant over the assimilation interval in virtue of the assumption $\mathscr{F}^\lambda = \mathbf{I}$. Equation (41) describes, in the linear approximation, the error evolution in the augmented dynamical system (40). The short-time approximation of the error dynamics (41) in the interval τ reads:

$$\delta\mathbf{z}^f \approx (\mathbf{M}\delta\mathbf{x}^a + \delta\boldsymbol{\mu}^a\tau, \delta\boldsymbol{\lambda}^a) . \tag{42}$$

As for the standard EKF, by taking the expectation of the product of (41) (or (42)) with its transpose, we obtain the forecast error covariance matrix, $\mathbf{P}_z^f$, for the augmented system:

$$\mathbf{P}_z^f = < \delta \mathbf{z}^f \delta \mathbf{z}^{fT} > = \begin{pmatrix} \mathbf{P}_x^f & \mathbf{P}_{x\lambda}^f \\ \mathbf{P}_{x\lambda}^{fT} & \mathbf{P}_\lambda^f \end{pmatrix} \tag{43}$$

where the $N \times N$ matrix $\mathbf{P}_x^f$ is the error covariance of the state estimate $\mathbf{x}^f$, $\mathbf{P}_\lambda^f$ is the $P \times P$ parametric error covariance and $\mathbf{P}_{x\lambda}^f$ the $N \times P$ error correlation matrix between the state vector, $\mathbf{x}$, and the vector of parameters $\boldsymbol{\lambda}$. These correlations are essential for the estimation of the parameters. In general one does not have access to a direct measurement of the parameters, and information are only obtained through observations of the system's state. As a consequence, at the analysis step, the estimate of the parameters will be updated only if they correlate with the system's state, that is $\mathbf{P}_{x\lambda}^f \neq 0$. The gain of information coming from the observations is thus spread out to the full augmented system phase space.

Let us define, in analogy with (43), the analysis error covariance matrix for the augmented system:

$$\mathbf{P}_z^a = \begin{pmatrix} \mathbf{P}_x^a & \mathbf{P}_{x\lambda}^a \\ \mathbf{P}_{x\lambda}^{aT} & \mathbf{P}_\lambda^a \end{pmatrix} \tag{44}$$

where the entries in (44) are defined as in (43) but refer now to the analysis step after the assimilation of observations.

By inserting (42) into (43), and taking the expectation, we obtain the forecast error covariance matrix in the linear and short-time approximation:

$$\mathbf{P}_x^f = \mathbf{M} < \delta \mathbf{x}^a \delta \mathbf{x}^{aT} > \mathbf{M}^T + < \delta \boldsymbol{\mu}^a \delta \boldsymbol{\mu}^{aT} > \tau^2 +$$

$$[\mathbf{M} < \delta \mathbf{x}^a \delta \boldsymbol{\mu}^{aT} > + < \delta \boldsymbol{\mu}^a \delta \mathbf{x}^{aT} > \mathbf{M}^T] \tau$$

$$= \mathbf{M} \mathbf{P}_x^a \mathbf{M}^T + \mathbf{Q}^a \tau^2 + [\mathbf{M} < \delta \mathbf{x}^a \delta \boldsymbol{\mu}^{aT} > + < \delta \boldsymbol{\mu}^a \delta \mathbf{x}^{aT} > \mathbf{M}^T] \tau \tag{45}$$

$$\mathbf{P}_\lambda^f = < \delta \boldsymbol{\lambda}^a \delta \boldsymbol{\lambda}^{aT} > \tag{46}$$

$$\mathbf{P}_{x\lambda}^f = \mathbf{M} < \delta \mathbf{x}^a \delta \boldsymbol{\lambda}^{aT} > + < \delta \boldsymbol{\mu}^a \delta \boldsymbol{\lambda}^{aT} > \tau \tag{47}$$

Note that (45) is equivalent to (32), except that now the correlations between the initial condition and the model error are maintained (last two terms on the r.h.s. of (45)), and $\mathbf{P}_x^a$ and $\mathbf{Q}^a$ replace $\mathbf{P}^a$ and $\mathbf{Q}$. Nevertheless, in contrast to the ST-EKF where $\mathbf{Q}$ is estimated statistically and then kept fixed, in the ST-AEKF $\mathbf{Q}^a$ is estimated online using the observations.

The information on the uncertainty in the model parameters is embedded in the error covariance $\mathbf{P}_\lambda^a$, a by-product of the assimilation. Using the definition of $\delta \boldsymbol{\mu}^a$

and (46), the matrix $\mathbf{Q}^a$ can be rewritten as:

$$\mathbf{Q}^a = <\delta\mu^a\delta\mu^{aT}> =$$

$$<\left(\frac{\partial\mathbf{f}}{\partial\lambda}|_{\lambda^a}\right)\delta\lambda^a\delta\lambda^{aT}\left(\frac{\partial\mathbf{f}}{\partial\lambda}|_{\lambda^a}\right)^T> \approx \left(\frac{\partial\mathbf{f}}{\partial\lambda}|_{\lambda^a}\right)\mathbf{P}_\lambda^a\left(\frac{\partial\mathbf{f}}{\partial\lambda}|_{\lambda^a}\right)^T. \qquad (48)$$

Similarly, the correlation terms in (45) can be written according to:

$$[\mathbf{M} <\delta\mathbf{x}^a\delta\mu^{aT}> + <\delta\mu^a\delta\mathbf{x}^{aT}> \mathbf{M}^T]\tau =$$

$$[\mathbf{M} <\delta\mathbf{x}^a\delta\lambda^{aT}\left(\frac{\partial\mathbf{f}}{\partial\lambda}|_{\lambda^a}\right)^T> + <\left(\frac{\partial\mathbf{f}}{\partial\lambda}|_{\lambda^a}\right)\delta\lambda^a\delta\mathbf{x}^{aT}> \mathbf{M}^T]\tau \approx$$

$$[\mathbf{M}\mathbf{P}_{\mathbf{x}\lambda}^a\left(\frac{\partial\mathbf{f}}{\partial\lambda}|_{\lambda^a}\right)^T + \left(\frac{\partial\mathbf{f}}{\partial\lambda}|_{\lambda^a}\right)\mathbf{P}_{\mathbf{x}\lambda}^{a\,T}\mathbf{M}^T]\tau. \qquad (49)$$

Using (48) and (49) in (45), the forecast state error covariance $\mathbf{P}_x^f$ can be written in terms of the state-augmented analysis error covariance matrix at the last observation time, according to:

$$\mathbf{P}_x^f \approx \mathbf{M}\mathbf{P}_x^a\mathbf{M}^T + \left(\frac{\partial\mathbf{f}}{\partial\lambda}|_{\lambda^a}\right)\mathbf{P}_\lambda^a\left(\frac{\partial\mathbf{f}}{\partial\lambda}|_{\lambda^a}\right)^T\tau^2 +$$

$$[\mathbf{M}\mathbf{P}_{\mathbf{x}\lambda}^a\left(\frac{\partial\mathbf{f}}{\partial\lambda}|_{\lambda^a}\right)^T + \left(\frac{\partial\mathbf{f}}{\partial\lambda}|_{\lambda^a}\right)\mathbf{P}_{\mathbf{x}\lambda}^{a\,T}\mathbf{M}^T]\tau. \qquad (50)$$

The three terms in (50) represent the contribution to the forecast state error covariance coming from the analysis error covariance in the system's state, in the parameters and in their correlation, respectively.

By making use of the definition of the model error vector $\delta\mu^a$ in (47), the forecast error correlation matrix $\mathbf{P}_{x\lambda}^f$ becomes:

$$\mathbf{P}_{x\lambda}^f \approx \mathbf{M}\mathbf{P}_{\mathbf{x}\lambda}^a + \left(\frac{\partial\mathbf{f}}{\partial\lambda}|_{\lambda^a}\right)\mathbf{P}_\lambda^a\tau. \qquad (51)$$

Expressions (46), (50) and (51) can be compacted into a single expression:

$$\mathbf{P}_z^f = \mathbf{C}\mathbf{P}_z^a\mathbf{C}^T \qquad (52)$$

with $\mathbf{C}$ being the ST-AEKF forward operator defined as:

$$\mathbf{C} = \begin{pmatrix} \mathbf{M} & \frac{\partial\mathbf{f}}{\partial\lambda}|_{\lambda^a}\tau \\ 0 & \mathbf{I}_P \end{pmatrix} \qquad (53)$$

where $\mathbf{I}_P$ is the $P \times P$ identity matrix.

The short-time bias Eq. (20) is used to estimate the bias in the state forecast, $\mathbf{x}^f$, due to parametric error, in analogy with the ST-EKF. This estimate is made online using the last innovation of the parameter vector. Assuming furthermore that the forecast of the parameter is unbiased, the bias in the state augmented forecast at time t_{k+1} reads:

$$\mathbf{b}_z^m = \begin{pmatrix} \mathbf{b}_x \\ \mathbf{b}_\lambda \end{pmatrix} = \begin{pmatrix} \frac{\partial \mathbf{f}}{\partial \lambda}|_{\lambda^a}(\lambda_{t_k}^a - \lambda_{t_k}^f)\tau \\ 0 \end{pmatrix}. \tag{54}$$

The bias $\mathbf{b}_z^m$ is then removed from the forecast field before the latter is used in the analysis update, that is $\tilde{\mathbf{z}}^f = \mathbf{z}^f - \mathbf{b}_z^m$, where $\tilde{\mathbf{z}}^f$ is the unbiased state augmented forecast.

As for the standard EKF, we need the observation operator linking the model to the observed variables. An augmented observation operator is introduced, $\mathscr{H}_z = [\mathscr{H} \quad 0]$ with $\mathscr{H}$ as in (23), and its linearization, $\mathbf{H}_z$ is now a $M \times (N+P)$ matrix in which the last P columns contain zeros; the rank deficiency in $\mathscr{H}$ reflects the lack of direct observations of the model parameters.

The augmented state and covariance update complete the algorithm:

$$\mathbf{z}^a = [\mathbf{I}_z - \mathbf{K}_z\mathbf{H}_z]\,\tilde{\mathbf{z}}^f + \mathbf{K}_z\mathbf{y}^o \tag{55}$$

$$\mathbf{P}_z^a = [\mathbf{I}_z - \mathbf{K}_z\mathbf{H}_z]\mathbf{P}_z^f \tag{56}$$

where the vector of observations $\mathbf{y}^o$ is the same as in the standard EKF while $\mathbf{I}_z$ is now the $(N+P) \times (N+P)$ identity matrix. The augmented gain matrix $\mathbf{K}_z$ is defined accordingly:

$$\mathbf{K}_z = \mathbf{P}_z^f\mathbf{H}_z^T \left[\mathbf{H}_z\mathbf{P}_z^f\mathbf{H}_z^T + \mathbf{R}\right]^{-1} \tag{57}$$

but it is now a $(I + P) \times M$ matrix.

Equations (40), (41), (42), (43), (44), (45), (46), (47), (48), (49), (50), (51) and (52) for the forecast step, and (55), (56) and (57) for the analysis update define the ST-AEKF. The algorithm is closed and self consistent meaning that, once it has been initialized, it does not need any external information (such as statistically estimated error covariances) and the state, the parameters and the associated error covariances are all estimated online using the observations.

The ST-AEKF is a short-time approximation of the classical augmented EKF, the AEKF [21]. In essence, the approximation consists of the use of an analytic expression for the evolution of the model error component of the forecast error covariance. This evolution law, quadratic for short-time, reflects a generic and intrinsic feature of the model error dynamics, connected to the model sensitivity to perturbed parameters and to the degree of dynamical instability. It does not depend on the specific numerical integration scheme adopted for the evolution of the model state. The state error covariance, $\mathbf{P}_x^f$, in the ST-AEKF, is evolved as in the standard AEKF: the propagator $\mathbf{M}$ is the product of the individual $\mathbf{M}_i$ relative to each time-step within the assimilation interval. The difference between the two algorithms is in the time propagation of the forecast error covariance associated with the misspecification of the parameters, $\mathbf{P}_{x\lambda}$. In the ST-AEKF this is reduced to the evaluation of the off diagonal term in the operator $\mathbf{C}$. This term replaces the full linearization of the model equations with respect to the estimated parameters, required by the AEKF. In this latter case the model equations are linearized with respect to the augmented state, $(\mathbf{x}, \lambda)$, giving rise to an augmented tangent linear model, $\mathbf{M}_z$. This linearization can be particularly involved [24], especially in the case of implicit or semi-implicit integration schemes such as those often used in NWP applications [23]. The propagator relative to the entire assimilation interval is then given by the product of the individual augmented tangent linear propagator over the single time-steps. As a consequence the cost of evolving the model error covariance in the AEKF grows with the assimilation interval. In the ST-AEKF, the use of the short-time approximation within the assimilation interval makes straightforward the implementation of the parameter estimation in the context of a pre-existing EKF, without the need to use an augmented tangent linear model during the data assimilation interval. It reduces the computational cost with respect to the AEKF, because the propagation of the model error component does not depend on the length of the assimilation interval. Nevertheless the simplifications in the setup and the reduction in the computational cost are obtained at the price of a decrease in the accuracy with respect to the AEKF. The degree of dynamical instabilities along with the length of the assimilation interval, are the key factors affecting the accuracy of the ST-AEKF.

3.2.1   Numerical Results with ST-EKF and ST-AEKF

Numerical experiments are carried out with two different models. OSSEs are performed first using the Lorenz [29] model used in Sect. 3.1.1, but in its one-scale version given by:

$$\frac{dx_i}{dt} = \alpha(x_{i+1} - x_{i-2})x_{i-1} - \beta x_i + F, \qquad i = \{1, \dots, 36\} \tag{58}$$

where the parameter associated with the advection, α, linear dissipation, β and the forcing F, are written explicitly. As for the experiments described in Sect. 3.1.1, the numerical integration are performed using a fourth-order Runge-Kutta scheme with a time step of 0.0083 units, corresponding to 1 h of simulated time.

The reference trajectory, representing the true evolution we intend to estimate, is given by a solution of (58) with parameters $\lambda^{tr} = (F^{tr}, \alpha^{tr}, \beta^{tr}) = (8, 1, 1)$; with this choice the model behaves chaotically. A network of $M = 18$ regularly spaced noisy observations is simulated by sampling the reference true trajectory and adding a Gaussian random observation error whose variance is set to 5 % of the system's climate variance. Model error is simulated by perturbing F, α and β with respect to their reference true values. Gaussian samples of 100 states and model parameters are used to initialize assimilation cycles lasting for 1 year. In all the experiments the initial condition error variance is set to 20 % of the system's climate variance. The model parameters are sampled from a Gaussian distribution with mean equal to λ^{tr} and standard deviation $\sigma_\lambda = 25\%$ of λ^{tr}.

We compare four configurations of the EKF: (1) standard EKF without model error treatment, (2) standard EKF using a perfect model, (3) ST-EKF (Sect. 3.1), and (4) ST-AEKF (Sect. 3.2). Recall that in the ST-EKF, model error bias and covariance are estimated according to (31) and (33) with $< \delta\mu_o >$ and Q evaluated on a statistical basis before the assimilation experiments. The expectation of $\delta\mu_o$ is estimated through:

$$< \delta\mu_o >=<< \frac{\partial \mathbf{f}}{\partial \lambda}|_\lambda (\lambda - \lambda^{tr}) >> \tag{59}$$

and is then used in (31) and (33). In (59) the averages are taken over the same Gaussian sample of initial conditions and parameters used to initialize the data assimilation experiments, using the actual value of the parameter, λ^{tr}, as the reference. This idealized procedure has been chosen to give the ST-EKF the best-possible statistical estimate of the model error in view of its comparison with the more sophisticated ST-AEKF.

Figure 3 shows the analysis error variance as a function of time for the four experiments of 1 year long; the assimilation interval is $\tau = 6$ h. The errors are

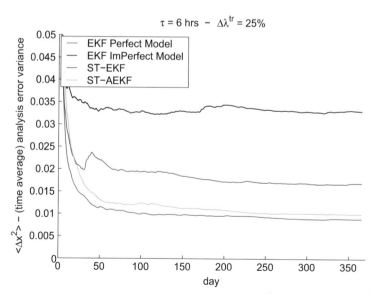

Fig. 3 Time averaged analysis error variance as a function of time. Standard EKF without model error treatment (*black*), standard EKF with perfect model (*red*), ST-EKF (*blue*) and ST-AEKF (*green*). The error variance is normalized with respect to the system's climate variance

spatio-temporal average over the ensemble of 100 experiments and over the model domain, and normalized with the system's climate variance. The figure clearly shows the advantage of incorporating a model error treatment: the average error of the ST-EKF is almost half of the corresponding to the standard EKF without model error treatment. However using the ST-AEKF the error is further reduced and attains a level very close to the perfect model case.

The benefit of incorporating a parameter estimation procedure in the ST-AEKF is displayed in Fig. 4 that shows the time mean analysis error variance for the EKF with perfect model and the ST-AEKF (top panel), along with the relative parametric errors as a function of time (bottom panel). The time series of the ST-AEKF analysis error variance is also superimposed to the time-averages in the top panel. Figure 4 reveals that the ST-AEKF is successful in recovering the true parameters. This reconstruction is very effective for the forcing, F, and the advection, α, and at a lesser extent for the dissipation, β. The ability of the ST-AEKF to efficiently exploit the observations of the system's state to estimate an uncertain parameter, either multiplicative or additive, is evident. Given that the innovation in the parameter, obtained via Eq. (55), is proportional to the cross-covariance forecast error, $\mathbf{P}^f_{x\lambda}$, the accuracy of the parameter estimation revealed by Fig. 4 turns out to be an indication of the quality of the short-time approximation, (51), on which the estimate of $\mathbf{P}^f_{x\lambda}$ is based.

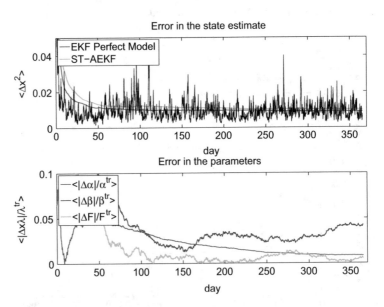

Fig. 4 *Top Panel* – Time averaged analysis error variance as a function of time: standard EKF with perfect model (*red*) and ST-AEKF (*green*); time series of the ST-AEKF (*black*). *Bottom Panel* – Absolute parametric error of the ST-AEKF, relative to the true value λ^{tr}. The error variance is normalized with respect to the system's climate variance

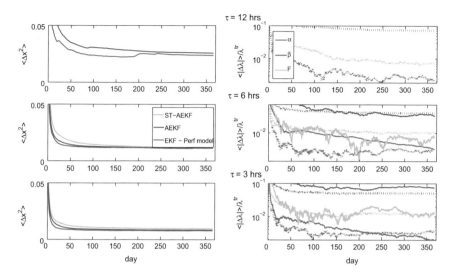

Fig. 5 *Left column*: Running mean of the quadratic state estimation error as a function of time; EKF with perfect model (*red*), ST-AEKF (*green*) and AEKF (*blue*). *Right column*: absolute value of the parametric error as a function of time for α (*red*), β (*blue*) and F (*green*), for ST-AEKF (*solid lines*) and AEKF (*dotted lines*). From *top* to *bottom* $\tau = 12$, 6 and 3 h respectively. The errors are averaged over an ensemble of 100 experiments and $\sigma_\lambda = 25\% \lambda^{tr}$

Figure 5 focuses on the comparison between the ST-AEKF and the standard AEKF. The experiments are carried out for $\tau = 3$, 6 and 12 h and with $\sigma_\lambda = 25\% \lambda^{tr}$. As above, the results are averaged over an ensemble of 100 experiments, and the observation error variance is 5 % of the system's climate variance. The left panels display the quadratic estimation error, while the parametric error is given in the panels in the right column; note that the logarithm scale is used in the y-axis. The estimation error relative to the EKF with a perfect model is also displayed for reference.

We see that as expected the AEKF has a superior skill than the ST-AEKF but for $\tau = 3$ or 6 h their performances are very similar. The AEKF shows a marked rapidity to reach convergence but the asymptotic error level attained by the two filters are practically indistinguishable. On the other hand for $\tau = 12$ h the ST-AEKF diverges whereas the AEKF is able to control error growth and maintain the estimation error to a low level. We first observe that in all but one cases the parametric error in the experiments with the AEKF is lower than for the ST-AEKF, in agreement with the observed lower state estimation error. However, when $\tau = 6$ or 3 h, the asymptotic parametric errors of the two filters are very similar, a remarkable result considering the approximate evolution law used in the ST-AEKF. An important difference is the extreme variability observed in the parametric error with the ST-AEKF as compared to the smoothness of the corresponding solutions with the AEKF. Note also that when $\tau = 6$ h the ST-AEKF reduces the error in the forcing more than the AEKF but the error curves are subject to very large fluctuations. The dissipation, β, appears as the most difficult parameter to be estimated in agreement with what observed in Fig. 4. In summary, Fig. 5 suggests that the ST-AEKF may represent a suitable and efficient alternative to the full AEKF when the assimilation interval does not exceed the time range of validity of the approximation on which the ST-AEKF is based. The results indicate that this limit is between 6 and 12 h given that the ST-AEKF diverges when $\tau = 12$ h. According to the theory outlined in [33], the short-time regime is related to the inverse of twice the largest (in absolute value) Lyapunov exponent of the system. In the Lorenz system (58) the largest Lyapunov exponent turns out to be the most negative one, equal to 0.97 day^{-1}, so that the duration of the short-time regime is estimated to be about 12 h, in qualitative agreement with the performance of the ST-AEKF. Finally note that the slight deterioration in the filter accuracy is compensated by a reduction in both the computational and implementation costs with respect to the AEKF.

The second model under consideration is an offline version of the operational soil model, Interactions between Surface, Biosphere,and Atmosphere (ISBA) [36]. In the experiments that follow, ST-AEKF has been implemented in the presence of parametric errors in the Leaf Area Index (LAI) and in the Albedo; more details, along with the case of other land surface parameters, can be found in [4]. OSSEs are performed using the two-layers version of ISBA which describes the evolution of soil temperature and moisture contents; the model is available within a surface externalized platform (SLDAS; [30]). The state vector, $\mathbf{x} = (T_s, T_2, w_g, w_2)$, contains the surface and deep soil temperatures T_s and T_2 and the corresponding water contents w_g and w_2. The vector $\boldsymbol{\lambda}$ is taken to represent the set of model parameters. A detailed description of ISBA can be found in [36].

The forcing data are the same for the truth and the assimilation solutions. They consist of 1-hourly air temperature, specific humidity, atmospheric pressure, incoming global radiation, incoming long-wave radiation, precipitation rate and wind speed relative to the ten summers in the decade 1990–1999 extract from ECMWF Re-analysis ERA40 and then dynamically down-scaled to 10 km horizontal resolution over Belgium [18]. The fields are then temporally interpolated to get data consistent with the time resolution of the integration scheme of ISBA (300 s). In this study ISBA is run in one offline single column mode for a 90 day period, and the forcing parameters are those relative to the grid point closest to Brussels. An one-point soil model has been also used by [37], for parameter estimation using an ensemble based assimilation algorithm.

The simulated observations are T_{2m} and RH_{2m}, interpolated between the forcing level (≈ 20 m) and the surface with the Geleyn's interpolation scheme ([16]), at 00, 06, 12 and 18 UTC. The assimilation interval is $\tau = 6$ h, while the observational noise is drawn from a Gaussian, $\mathcal{N}(0, \mathbf{R})$, with zero-mean and covariance given by the diagonal matrix $\mathbf{R}$ with elements: $diag(\mathbf{R}) = (\sigma_{T_{2m}}^2, \sigma_{w_{2m}}^2) = (1K^2, 10^{-2})$. As explained in [31], the observation operator $\mathcal{H}$, relating the state vector to the observation includes the model integration. The initial $\mathbf{P}^a$ and $\mathbf{P}^m$ required by the EKF are set as diagonal with elements $diag(\mathbf{P}^a) = (\sigma_{T_s}^2, \sigma_{T_2}^2, \sigma_{w_g}^2, \sigma_{w_2}^2) =$

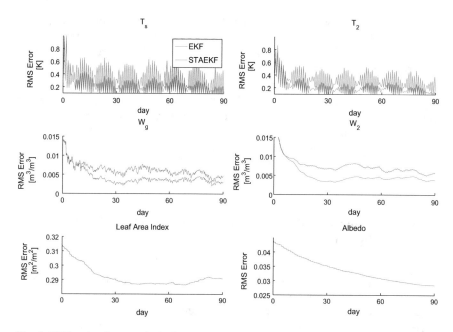

Fig. 6 RMS estimation error in the four state variables for the EKF (*red*) and the STAEKF (*blue*). The RMS error in the estimate of *LAI* and Albedo relative to the ST-AEKF are shown in the *bottom-left/right panels* respectively (Modified from [4])

$(1K^2, 1K^2, 10^{-2}, 10^{-2})^1$ and $diag(\mathbf{P}^m) = (\sigma_{T_s}^2, \sigma_{T_2}^2, \sigma_{w_g}^2, \sigma_{w_2}^2) = (25 \times 10^{-2}K^2, 25 \times 10^{-2}K^2, 4 \times 10^{-4}, 4 \times 10^{-4})$ [30].

Parametric errors is introduced by perturbing simultaneously the LAI and the albedo. These parameters strongly influence the surface energy balance budget and partitioning, which in turn regulate the circulation patterns and modify the hydrological processes. For each summer in the period 1990–1999, a reference trajectory is generated by integrating the model with LAI $= 1\,\text{m}^2/\text{m}^2$ and albedo $= 0.2$. Around each of these trajectories, Gaussian samples of 100 initial conditions and uncertain parameters are used to initialize the assimilation cycles. The initial conditions are sampled from a distribution with standard deviation $(\sigma_{T_s}, \sigma_{T_2}, \sigma_{w_g}, \sigma_{w_2}) = (5K, 5K, 1, 1)$, whereas LAI and the albedo are sampled with standard deviations, $\sigma_{LAI} = 0.5\,\text{m}^2/\text{m}^2$ and $\sigma_{albedo} = 0.05$ respectively ([17]). The initial $\mathbf{P}_\lambda^a$ in the ST-AEKF read $\mathbf{P}_{LAI}^a = 1\,(\text{m}^2/\text{m}^2)^2$ and $\mathbf{P}_{albedo}^a = 10^{-4}$, $\mathbf{P}_\mathbf{x}^a$ is taken as in the EKF while $\mathbf{P}_{\mathbf{x},\lambda}^a$ is initially set to zero.

Results are summarized in Fig. 6 which shows the RMS Error in the four state variable for the EKF and ST-AEKF, along with the RMS Error in LAI and Albedo for the ST-AEKF. The progressive parametric error reduction achieved

---

[1]The values of σ_{w_g} and σ_{w_2} are expressed as soil wetness index $SWI = (w - w_{wilt})/(w_{fc} - w_{wilt})$ where w_{fc} is the volumetric field capacity and w_{wilt} is the wilting point.

with the ST-AEKF is reflected by the systematically lower estimation error in the soil temperature and water content. At the very initial times, of the order of 1 week, EKF and ST-AEKF have an indistinguishable skill. However, as soon as the state-parameter error correlations in the ST-AEKF augmented forecast error matrix become mature, the improvement of the ST-AEKF becomes apparent and it lasts for the entire duration of the experiment. By reducing the parametric error, a better guess for the system state can be obtained and this in turn improves the analysis field and again the accuracy of the parameter estimate. Moreover, given that this feature is incorporated using the short-time formulation [6], the additional computational cost with respect to the standard EKF is almost negligible.

4   Deterministic Model Error Treatment in Variational Data Assimilation

Variational assimilation attempts to solve the smoothing problem of a simultaneous fit to all observations distributed within a given interval of interest. We suppose therefore that I measurements, (23), are collected at the discrete times, $(t_1, t_2, \ldots, t_I)$, within a reference time interval T. An priori estimation, $\mathbf{x}_b$, of the model initial condition is supposed to be available. This is usually referred to as the background state, and:

$$\mathbf{x}_0 = \mathbf{x}_b + \boldsymbol{\varepsilon}_b \tag{60}$$

where $\boldsymbol{\varepsilon}_b$ represents the background error.

We search for the trajectory that, on the basis of the background field and according to some specified criteria, best fits the observations over the reference period T. Besides the observations and the background, the model dynamics itself represents a source of additional information to be exploited in the state estimate. The model is not perfect, and we assume that an additive error affects the model prediction in the form:

$$\mathbf{x}(t) = \mathcal{M}(\mathbf{x}_0) + \delta\mathbf{x}^m(t) . \tag{61}$$

Assuming furthermore that all errors are Gaussian and do not correlate with each other, the quadratic penalty functional, combining all the information, takes the form [21]:

$$2J = \int_0^T \int_0^T (\delta\mathbf{x}^m(t'))^T (\mathbf{P}^m)^{-1}_{t't''} (\delta\mathbf{x}^m(t'')) dt' \, dt'' +$$

$$\sum_{k=1}^I (\boldsymbol{\varepsilon}_k^o)^T \mathbf{R}_k^{-1} (\boldsymbol{\varepsilon}_k^o) + \boldsymbol{\varepsilon}_b^T \mathbf{B}^{-1} \boldsymbol{\varepsilon}_b . \tag{62}$$

The weighting matrices $\mathbf{P}_{t't''} = \mathbf{P}(t', t'')$, $\mathbf{R}_k$ and $\mathbf{B}$ have to be regarded as a measure of our confidence in the model, in the observations and in the background field, respectively. In this Gaussian formulation these weights can be chosen to reflect the relevant moments of the corresponding Gaussian error distributions. The best-fit is defined as the solution, $\hat{\mathbf{x}}(t)$, minimizing the cost-function J over the interval T. It is known that, under the aforementioned hypothesis of Gaussian errors, $\hat{\mathbf{x}}(t)$ corresponds to the maximum likelihood solution and J can be used to define a multivariate distribution of $\mathbf{x}(t)$ [21]. Note that, in order to minimize J all errors have to be explicitly written as a function of the trajectory $\mathbf{x}(t)$.

The variational problem defined by (62) is usually referred to as *weak-constraint* given that the model dynamics is affected by errors [40]. An important particular case is the *strong-constraint* variational assimilation in which the model is assumed to be perfect, that is $\delta\mathbf{x}^m = 0$, [13, 27]. In this case the model-error related term disappears and the cost-function reads:

$$2J_{strong} = \sum_{k=1}^{I} (\boldsymbol{\varepsilon}_k^o)^T \mathbf{R}_k^{-1} (\boldsymbol{\varepsilon}_k^o) + \boldsymbol{\varepsilon}_b^T \mathbf{B}^{-1} \boldsymbol{\varepsilon}_b . \tag{63}$$

The calculus of variations can be used to find the extremum of (62) (or (63)) and leads to the corresponding Euler-Lagrange equations [3, 13]. In the strong-constraint case, the requirement that the solution has to follow the dynamics exactly is satisfied by appending to (63) the model equations as a constraint by using a proper Lagrange multiplier field. However the size and complexity of the typical NWP problems is such that the Euler-Lagrange equations cannot be practically solved unless drastic approximations are introduced. When the dynamics is linear and the amount of observations is not very large, the Euler-Lagrange equations can be efficiently solved with the method of representers [3]. An extension of this approach to nonlinear dynamics has been proposed in [44]. Nevertheless, the representers method is far from being applicable for realistic high dimensional problems, like the numerical weather prediction and an attractive alternative is represented by the descent methods which makes use of the gradient vector of the cost-function in an iterative minimization procedure [41]. This latter approach is used in most of the operational NWP centers which employ variational assimilation. Note that in the cost-functions (62) model error is allowed to be correlated in time, and gives up the double integral in the first r.h.s. term. If model error is assumed to be a random uncorrelated noise, only covariances have to be taken into account and the double integral reduces to a single integral (to a single summation in the discrete time case).

The search for the best-fit trajectory by minimizing the associated cost-function requires the specification of the weighting matrices. The estimation of the matrices $\mathbf{P}_{t't'}^m$ is particularly difficult in realistic NWP applications due to the large size of the typical models currently in use. Therefore it becomes crucial to define approaches for modeling the matrices $\mathbf{P}_{t't'}^m$ and reduce the number of parameters required for their estimation. We will show below how the deterministic, and short-time, model error formulation described in Sect. 2.2 can be used to derive $\mathbf{P}^m_{t't''}$

We make the conjecture that, as long as the errors in the initial conditions and in the model parameters are small, the second rhs term of (12), $\delta \mathbf{x}^m(t) = \int_{t_0}^t d\tau \mathbf{M}_{t,\tau} \delta \mu(\tau)$ can be used to estimate the model error entering the weak-constraint cost-function, and the corresponding correlation matrices $\mathbf{P}^m(t', t'')$. In this case, the model error dependence on the model state, induces the dependence of model error correlation on the correlation time scale of the model variables themselves. By taking the expectation of the product of the second rhs term of (12) by itself, over an ensemble of realizations around a specific trajectory, we obtain an equation for the model error correlation matrix:

$$\mathbf{P}^m(t', t'') = \int_{t_0}^{t'} d\tau \int_{t_0}^{t''} d\tau' \mathbf{M}_{t',\tau} < \delta \mu(\tau) \delta \mu(\tau')^T > \mathbf{M}^T_{t'',\tau'} . \qquad (64)$$

The integral equation (64) gives the model error correlation between times t' and t''. In this form, Eq. (64) is of little practical use for any realistic non-linear systems. A suitable expression can be obtained by considering its short-time approximation through a Taylor expansion around $(t', t'') = (t_0, t_0)$. It can be shown [5] that the first non-trivial order is quadratic and reads:

$$\mathbf{P}(t', t'') \approx < \delta \mu_0 \delta \mu_0^T > (t' - t_0)(t'' - t_0) . \qquad (65)$$

Equation (65) states that the model error correlation between two arbitrary times, t' and t'', within the short-time regime, is equal to the model error covariance at the origin, $< \delta \mu_0 \delta \mu_0^T >$, multiplied by the product of the two time intervals. Naturally the accuracy of this approximation is connected on the one hand to the length of the reference time period, on the other to the accuracy of the knowledge about the error in the parameters needed to estimate $< \delta \mu_0 \delta \mu_0^T >$. We propose to use the short-time relation (65) as an estimate of the model error correlations in the variational assimilation. The resulting algorithm is hereafter referred to as Short-Time-Weak-Constraint-4DVar (ST-w4DVar). Besides the fact of being a short-time approximation, (65) is based on the hypothesis that the error dynamics can be linearised. To highlight advantages and drawbacks of its application, we explicitly compare ST-w4DVar with other formulations.

4.1 Numerical Results with ST-w4DVar

The analysis is carried out in the context of two systems of increasing complexity. We first deal with a very simple example of scalar dynamics which is fully integrable. The variational problem is solved with the technique of representers. The simplicity of the dynamics allows us to explicitly solve (64) and use it to estimate the model error correlations. This "full weak-constraint" formulation of the 4DVar is evaluated and compared with the ST-w4DVar employing the short-time evolution law (65). In addition, a comparison is made with the widely used strong-constraint 4DVar in which the model is considered as perfect. In the last part of the Section we extend the analysis to an idealized nonlinear chaotic system. In this case the minimization is made by using an iterative descent method which makes use of the cost-function gradient. In this nonlinear context ST-w4DVar is compared to the strong-constraint and to a weak-constraint 4DVar in which model error is treated as a random uncorrelated noise as it is often assumed in realistic applications.

Let us consider the simple scalar dynamics:

$$x(t) = x_0 e^{\lambda^{tr} t} \tag{66}$$

with $\lambda^{tr} > 0$, as our reference.

Suppose that I noisy observations of the state variable are available at the discrete times $t_k \in [0, T]$, $1 \le k \le I$:

$$y_k^o = x_k + \varepsilon_k^o$$

ε_k^o being an additive random noise with variance $\sigma_o^2(t_k) = \sigma_o^2$, $1 \le k \le I$, and that a background estimate, x_b, of the initial condition, x_0, is at our disposal:

$$x_0 = x_b + \varepsilon_b$$

with ε_b being the background error with variance σ_b^2. We assume the model is given by:

$$x(t) = x_0 e^{\lambda t}.$$

We seek for a solution minimizing simultaneously the error associated with all these information sources. The quadratic cost function can be written in this case as:

$$2J(x) = \int_0^T \int_0^T (x(t') - x_0 e^{\lambda t'}) p_{t't''}^{-2} (x(t'') - x_0 e^{\lambda t''}) dt' dt'' +$$

$$\sum_{k=1}^I \sigma_o^{-2}(y_k^o - x_k)^2 + \sigma_b^{-2}(x_0 - x_b)^2 . \tag{67}$$

The control variable here is the entire trajectory within the assimilation interval T. In Eq. (67) we have used the fact that the model error bias, $\delta x^m(t)$, is given by $x(t) - x_0 e^{\lambda t}$ assuming the model and the control trajectory, $x(t)$, are started from the same initial condition x_0. Note that x_0 is itself part of the estimation problem through the background term in the cost-function, and that the covariance matrices all reduce to scalar, such as $p_{t't''} = p(t', t'')$.

While complete details can be found in [5] we describe here the essential of the derivation. The final minimizing solution of (67) is found using the technique of representer and reads:

$$x(t) = x_b e^{\lambda t} + \sum_{k=1}^{I} \beta_k r_k(t) = x^f(t) + \sum_{k=1}^{I} \beta_k r_k(t) \qquad 0 \le t \le T. \qquad (68)$$

The I functions, $r_k(t)$, are the representers given by:

$$r_k(t) = r_k(0)e^{\lambda t} + \int_0^T p_{tt'}^2(t')a_k(t')dt' \qquad 1 \le k \le I \qquad (69)$$

subject to $r_k(0) = \sigma_b^2 \int_0^T a_k(t)e^{\lambda t}dt$, $1 \le k \le I$, while the adjoint representers satisfy:

$$a_k(t) = \delta(t - t_k) \qquad 1 \le k \le I \qquad (70)$$

subject to $a_k(T) = 0$, $1 \le k \le I$. The coefficients, β_k, are given by:

$$\boldsymbol{\beta} = (\mathbf{S} + \sigma_o^2 \mathbf{I}_d)^{-1} \mathbf{d} \qquad (71)$$

with $\mathbf{d}$ the innovation vector, $\mathbf{d} = (y_1^o - x_1^f, \ldots, y_I^o - x_I^f)$, $\mathbf{S}$ the $I \times I$ matrix $(\mathbf{S})_{i,j} = r_i(t_j)$, and $\mathbf{I}_d$ the $I \times I$ identity matrix. The coefficients are then inserted in (68) to obtain the final solution.

In the derivation of the general solution (68) (with the coefficients (71)), we have not specified the model error correlations $p^2(t', t'')$; the particular choice adopted characterizes the formulations we aim to compare. Our first choice consists in evaluating the model error correlations through (64). By inserting $\delta\mu = \frac{\partial f}{\partial \lambda}\delta\lambda$, with $f(x) = \lambda x$, and the fundamental matrix, $\mathbf{M}_{t,t_0} = e^{\lambda(t-t_0)}$, associated with the dynamics (66), we get:

$$p^2(t', t'') = <(x_0\delta\lambda)^2 > e^{\lambda(t'+t'')}t't'' \qquad (72)$$

where the expectation, $<>$, is an average over a sample of initial conditions and parametric errors. Expression (72) can now be inserted into (69)–(70), to obtain the I representer functions:

$$r_k(t) = e^{\lambda(t+t_k)}[<(x_0\delta\lambda)^2 > t_k t + \sigma_b^2] \qquad 1 \le k \le M. \qquad (73)$$

The representers (73) are then inserted into (71) to obtain the coefficients for the solution, $x(t)$, which is finally obtained through (68). This solution is hereafter referred to as the full weak-constraint.

The same derivation is now repeated with the model error weights given by the short-time approximation (65). By substituting $\delta\mu = \frac{\partial f}{\partial\lambda}\delta\lambda$ into (65), we obtain:

$$p^2(t', t'') =< (x_0\delta\lambda)^2 > t' t'' . \tag{74}$$

Once (74) is inserted into (69)–(70) the representer solutions become:

$$r_k(t) = \sigma_b^2 e^{\lambda(t+t_k)} + < (x_0\delta\lambda)^2 > t_k t \qquad 1 \leq k \leq I . \tag{75}$$

The representer functions are then introduced into (71) and (68) to obtain the solution, $x(t)$, during the reference period T. The solution based on (75) is the ST-w4DVar.

The strong-constraint solution is derived by invoking the continuity of the solution (73), or (75), with respect to the model error weights. The strong-constraint solution is obtained in the limit $\delta\lambda \to 0$, and reads:

$$r_k(t) = \sigma_b^2 e^{\lambda(t+t_k)} \qquad 1 \leq k \leq I . \tag{76}$$

The three solutions based respectively on (73), (75) and (76) are compared in Fig. 7. Simulated noisy observations sampled from a Gaussian distribution around a solution of (66), are distributed every 5 time units over an assimilation interval $T = 50$ time units. Different regimes of motion are considered by varying the true parameter λ^{tr}. The results displayed in Fig. 7 are averages over 10^3 initial conditions and parametric model errors, around $x_0 = 2$ and λ^{tr}, respectively. The initial conditions are sampled from a Gaussian distribution with standard deviation $\sigma_b = 1$, while the model parameter, λ, is sampled by a Gaussian distribution with standard deviation $|\Delta\lambda| = |\lambda^{tr} - \lambda|$; the observation error standard deviation is $\sigma_o = 0.5$.

Figure 7 shows the mean quadratic estimation error, as a function of time, during the assimilation period T. The different panels refer to experiments with different parameter for the truth $0.01 \leq \lambda^{tr} \leq 0.03$, while the parametric error relative to the true value is set to $\Delta\lambda/\lambda^{tr} = 50\%$. The three lines refer to the full weak-constraint (dashed line), ST-w4DVar (continuous line) and the strong-constraint (dotted line) solutions respectively. The bottom right panel summarizes the results and shows the mean error, averaged also in time, as a function of λ^{tr} for the weak-constraint solutions only. As expected the full weak-constraint solution performs systematically better than any other approach. ST-w4DVar successfully outperforms the strong-constraint case, particularly at the beginning and end of the assimilation interval. The last plot displays the increase of total error of this solution as a function of λ^{tr}. To understand this dependence, one must recall that the duration of the short-time regime in a chaotic system is bounded by the inverse of the largest amplitude Lyapunov exponent [33]. For the scalar unstable case considered here, this role is played by the parameter λ^{tr}. The increase of the total error of the short-time approximated weak-constraint as a function of λ^{tr} reflects the progressive decrease

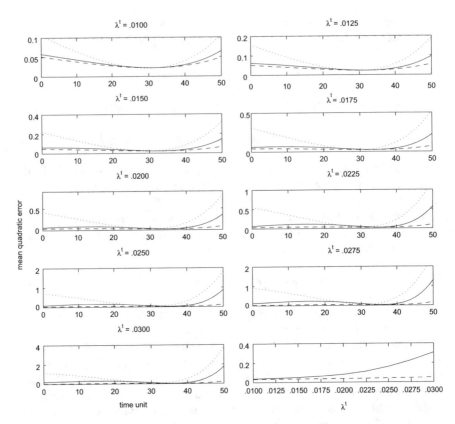

Fig. 7 Mean quadratic estimation error as a function of time for variational assimilation with system (66). The panels refer to experiments with different λ^{tr}. The *bottom-right panel* shows the mean quadratic error, for the weak-constraint solutions only, averaged also over the assimilation interval T as a function of λ^{tr}. Strong-constraint solution (*dotted line*), full weak-constraint solution (*dashed line*), short-time approximated weak-constraint solution (*continuous line*) (From [5] ©American Meteorological Society; used with permission)

of the accuracy of the short-time approximation for this fixed data assimilation interval, T.

The accuracy of the ST-w4DVar in relation to the level of instability of the dynamics, is further summarized in Fig. 8, where the difference between the mean quadratic error of this solution and the full weak-constraint one, is plotted as a function of the adimensional parameter $T\lambda^{tr}$, with $10 \leq T \leq 60$ and $0.0100 \leq \lambda^{tr} \leq 0.0275$. In all the experiments $\Delta\lambda/\lambda^{tr} = 50\%$. Remarkably all curves are superimposed, a clear indication that the accuracy of the analysis depends essentially on the product of the instability of the system and the data assimilation interval.

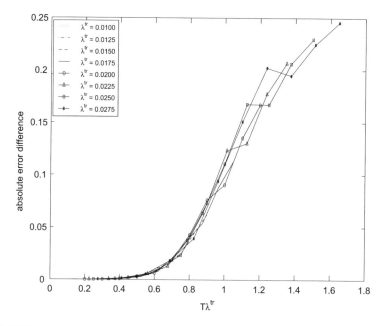

Fig. 8 Difference between the mean quadratic error of the short-time approximated and the full weak-constraint solution, for system (66), as a function of $T\lambda^{tr}$, for assimilation period $10 \leq T \leq 60$ and for different values of λ^{tr}. In all the experiments $\Delta\lambda/\lambda^{tr} = 50\%$ (From [5] ©American Meteorological Society; used with permission)

We turn now to the case of a nonlinear dynamics. We adopt here the widely used Lorenz 3-variable convective system [28], whose equations read:

$$\frac{dx}{dt} = -\sigma(x - y)$$

$$\frac{dy}{dt} = \rho x - y - xz \qquad (77)$$

$$\frac{dz}{dt} = xy - \beta z$$

with $\lambda = (\sigma, \rho, \beta) = (10, 28, \frac{8}{3})$. OSSEs are performed with a solution of (77) representing the reference dynamics from which observations are sampled. The estimation is based on observations of the entire system's state (i.e. observation operator equal to the identity 3×3 matrix), distributed within a given assimilation interval and affected by an uncorrelated Gaussian error with covariance **R**. The model dynamics is given by (77) with a modified set of parameters. The numerical integrations are carried out with a second order Runge-Kutta scheme with a time-step equal to 0.01 adimensional time units.

The variational cost function can be written, according to (62), as:

$$2J(\mathbf{x}_0, \mathbf{x}_1, \ldots, \mathbf{x}_L) = \sum_{i=1}^{L} \sum_{j=1}^{L} (\mathbf{x}_i - \mathcal{M}(\mathbf{x}_{i-1}))^T (\mathbf{P}^m)_{i,j}^{-1} (\mathbf{x}_j - \mathcal{M}(\mathbf{x}_{j-1})) +$$

$$\sum_{k=1}^{I} (\mathbf{y}_k^o - \mathcal{H}(\mathbf{x}_k))^T \mathbf{R}^{-1} (\mathbf{y}_k^o - \mathcal{H}(\mathbf{x}_k)) + (\mathbf{x}_b - \mathbf{x}_0)^T \mathbf{B}^{-1} (\mathbf{x}_b - \mathbf{x}_0) . \qquad (78)$$

We have assumed the assimilation interval T has been discretized over L time steps of equal length, Δt.

The control variable for the minimization is the series of the model state $\mathbf{x}_i$ at each time-step in the interval T. The minimizing solution is obtained by using a descent iterative method which makes use of the cost-function gradient with respect to $\mathbf{x}_i$, $0 \leq i \leq L$. This latter reads:

$$\nabla_{\mathbf{x}_0} J = -\mathbf{H}_0^T \mathbf{R}^{-1} (\mathbf{y}_0^o - \mathcal{H}(\mathbf{x}_0))$$

$$-\mathbf{M}_{0,1}^T [\sum_{j=1}^{L} (\mathbf{P}_{1,j}^m)^{-1} (\mathbf{x}_j - \mathcal{M}(\mathbf{x}_{j-1}))] - \mathbf{B}^{-1}(\mathbf{x}_b - \mathbf{x}_0) \qquad i = 0$$

$$\nabla_{\mathbf{x}_i} J = -\mathbf{H}_i^T \mathbf{R}^{-1} (\mathbf{y}_i^o - \mathcal{H}(\mathbf{x}_i)) - \mathbf{M}_{i,i+1}^T [\sum_{j=1}^{L} (\mathbf{P}_{i+1,j}^m)^{-1} (\mathbf{x}_j - \mathcal{M}(\mathbf{x}_{j-1}))]$$

$$+ \sum_{j=1}^{L} \mathbf{P}_{i,j}^{-1} (\mathbf{x}_j - \mathcal{M}(\mathbf{x}_{j-1})) \qquad 1 \leq i \leq L - 1 \qquad (79)$$

$$\nabla_{\mathbf{x}_L} J = -\mathbf{H}_L^T \mathbf{R}^{-1} (\mathbf{y}_L^o - \mathcal{H}(\mathbf{x}_L)) + \sum_{j=1}^{L} (\mathbf{P}_{L,j}^m)^{-1} (\mathbf{x}_j - \mathcal{M}(\mathbf{x}_{j-1})) \qquad i = L .$$

The gradient (79) is derived assuming that observations are available at each time step t_i, $0 \leq i \leq L$. In the usual case of sparse observations the term proportional to the innovation disappears from the gradient with respect to the state vector at a time when observations are not present. Note furthermore that, if the model error is treated as an uncorrelated noise, the corresponding term in the cost-function reduces to a single summation over the time-steps weighted by the inverse of the model error covariances. The cost-function gradient modifies accordingly and the summation over all time-steps disappears [42].

The cost-function (78) and its gradient (79) define the discrete weak-constraint variational problem. The ST-w4DVar consists in (78) and (79) with the model error correlations $\mathbf{P}_{i,j}^m$ estimated using the short-time approximation (65) that, in this discrete case, reads:

$$\mathbf{P}_{i,j}^m = < \delta\boldsymbol{\mu}_0 \delta\boldsymbol{\mu}_0^T > ij\Delta t^2 . \qquad (80)$$

The invariant term $< \delta\boldsymbol{\mu}_0\delta\boldsymbol{\mu}_0^T >$, which is here a 3×3 symmetric matrix, is assumed known a-priori and estimated by accumulating statistics on the model attractor, so that $< \delta\boldsymbol{\mu}_0\delta\boldsymbol{\mu}_0^T > = << \delta\boldsymbol{\mu}_0\delta\boldsymbol{\mu}_0^T >>$ and perturbing randomly each of the three parameters σ, ρ and β, with respect to the canonical values and with a standard deviation $|\Delta\boldsymbol{\lambda}|$. The ST-w4DVar is compared with the weak-contraint 4DVar with uncorrelated model error formulation and with the strong-constraint 4DVar; in this latter case the model error term disappears from the cost-function (78) and the gradient is computed with respect to the initial condition only [41].

The assumption of uncorrelated model error is done often in applications. It is particularly attractive in view of the important reduction of the computational cost associated with the minimization procedure. Model error covariance are commonly modeled as proportional to the background matrix, so that $\mathbf{P}^m = \alpha\mathbf{B}$. Figure 9 shows the mean quadratic error as a function of the tuning parameter. Results are averaged over an ensemble of 50 initial conditions and parametric model error; the observation and assimilation interval are set to 2 and 8 time-steps respectively. Strong constraint 4DVar (green) and ST-w4DVar (red) do not depend on α and are therefore horizontal lines in the panel. Weak constraint 4DVar with uncorrelated model error (blue) shows, as expected, a marked dependence on the model error covariance amplitude. The blue line with squared marks refers to an experiment where the model error is treated as an uncorrelated noise but the spatial covariances at observing times are estimated using the short-time approximation. Comparing this curve with the blue line relative to weak constraint 4DVar with uncorrelated model error and $\mathbf{P}^m = \alpha\mathbf{B}$ allow for evaluating the impact of neglecting the time correlation and of using an incorrect spatial covariance.

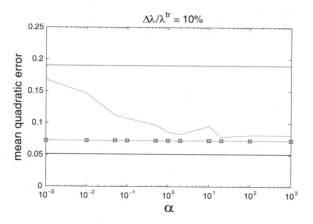

Fig. 9 Mean quadratic estimation error as a function of the tuning parameter α multiplying the model error covariance in the weak-constraint 4DVar with the uncorrelated noise assumption (see text for details). The dynamics is given by system (77). ST-w4DVar (*red*), strong-constraint 4DVar (*green*), uncorrelated noise weak-constraint 4DVar (*blue*) and uncorrelated noise weak-constraint 4DVar with spatial covariance as in the short-time approximated weak-constraint (*blue* with *red marks*)

The uncorrelated noise formulation (solid line with no marks) never reaches the accuracy of the ST-w4DVar. Note furthermore that for small α it almost reaches the same error level as the strong-constraint 4DVar where the model is assumed to be perfect. By further increasing α over $\alpha = 10^3$ (not shown) the error reaches a plateau whose value is controlled by the observation error level. When the spatial covariances are estimated as in the short-time weak-constraint the performance is generally improved, although for large α, the estimate $\mathbf{P}^m = \alpha\mathbf{B}$ gives very close skill and the improvement in correspondence with the best-possible α is only minor. This suggests that the degradation of the uncorrelated noise formulation over the short-time weak-constraint is mainly the consequence of neglecting the time correlation and only to a small extent to the use of an incorrect spatial covariance.

5 Discussion

Data assimilation schemes are usually assuming the uncorrelated nature of model uncertainties. This choice is indeed legitimate as a first approximation when initial condition errors dominate model errors. Due to the large increase of measurement data availability and quality, this view should be reassessed. However one prominent difficulty in dealing with model errors is the wide variety of potential sources of uncertainties, going from parametric errors up to the absence of description of some dynamical processes. But recently a stream of works, [33–35], provided important insights into the dynamics of deterministic model uncertainties, and from which generic mechanisms of growth were disentangled. In particular, it was shown that the mean square error associated with the presence of deterministic model errors is growing quadratically in time at short lead times. These insights now open new avenues in the description of data assimilation schemes, as it has been demonstrated in the present chapter.

First, the deterministic approach was applied in the context of the Extended Kalman Filter, for both the classical state estimation scheme and its augmented version (state and parameters). It has been demonstrated that these new schemes are indeed most valuable when dealing with deterministic model error sources. They provide a large improvement in the state estimation (in particular with the ST-AEKF) not only in the context of idealized settings (Lorenz' system) but also for realistic applications as shown by the results obtained with the offline version of an operational soil model (ISBA). Second the same idea was investigated in the context of four dimensional variational assimilation for which a weak-constrained framework was adopted. In this case model error cross-correlations were considered as quantities depending quadratically on time, implying a time dependent weighting of the model error terms during the assimilation period. This approach also led to important improvements as compared to more drastic assumptions like the time independence of model error source terms, or the absence of such sources.

The approaches proposed in this chapter rely on an important assumption, the deterministic nature of model uncertainties. As alluded in Sect. 2, the proposed general setting could also be extended to independent random noises, provided the appropriate temporal variations of the bias (Eq. 5) and covariances (Eq. 7) are used (see the remark at the end of Sect. 2). Still the covariance will be dependent on time (linearly) (see e.g. [45]) and will for instance affect differently the weights of the weak constrained four dimensional variational assimilation. Besides these technical aspects, it remains difficult to know what the exact nature of the sources of model errors is. A realistic view would be that the fast time scale processes – like turbulence in the surface boundary layer – could be considered as random components, while parameterization errors in the larger scale description of the stability of the atmosphere is a deterministic process. This leads us to consider model errors at a certain scale of description as a deterministic component plus a random component, and the user is left with the delicate question of evaluating whether the dominant sources present in his problem are of one kind or the other.

This work has mostly tackled this problem in the context of simple idealized dynamical systems. These encouraging results need however to be confronted with more operational problems. In line with the results obtained with soil model, this problem is currently under investigation in the context of the use of the ST-AEKF for an online version of ISBA coupled with ALARO at the Royal Meteorological Institute of Belgium. A relevant methodological development will be the incorporation of the deterministic model error treatment in the context of the ensemble based schemes, as illustrated in [32] or [38]. Finally, it is worth mentioning that the deterministic model error dynamics has been recently used in a new drift correction procedure in the context of interannual-to-decadal predictions [9]. These recent applications illustrate the usefulness of the deterministic approach and should be further extended to a wider range of applications in which model error is present. In particular in the context of coupled atmosphere-ocean systems where multiple scales of motion are present and model error often originates at the level of the coupling.

References

1. Anderson, J.L., Anderson, S.L.: A Monte Carlo implementation of the nonlinear filtering problem to produce ensemble assimilations and forecasts. Mon. Weather Rev. **127**(12), 2741–2758 (1999)
2. Bengtsson, L., Ghil, M., Kallen, E.: Dynamic Meteorology: Data Assimilation Methods. Springer, New York/Heidelberg/Berlin (1981)
3. Bennett, A.F.: Inverse Methods in Physical Oceanography. Cambridge University Press, Cambridge (1992)
4. Carrassi, A., Hamdi, R., Termonia, P., Vannitsem, S.: Short time augmented extended Kalman filter for soil analysis: a feasibility study. Atmos. Sci. Lett. **13**, 268–274 (2012)
5. Carrassi, A., Vannitsem, S.: Accounting for model error in variational data assimilation: a deterministic formulation. Mon. Weather Rev. **138**, 3369–3386 (2010)
6. Carrassi, A., Vannitsem, S.: State and parameter estimation with the extended Kalman filter: an alternative formulation of the model error dynamics. Q. J. R. Meteor. Soc. **137**, 435–451 (2011)

7. Carrassi, A., Vannitsem, S.: Treatment of the error due to unresolved scales in sequential data assimilation. Int. J. Bifurc. Chaos. **21**(12), 3619–3626 (2011)
8. Carrassi, A., Vannitsem, S., Nicolis, C.: Model error and sequential data assimilation: a deterministic formulation. Q. J. R. Meteor. Soc. **134**, 1297–1313 (2008)
9. Carrassi, A., Weber, R., Guemas, V., Doblas-Reyes, F., Asif, M., Volpi, D.: Full-field and anomaly initialization using a low-order climate model: a comparison and proposals for advanced formulations. Nonlinear Proc. Geophys. **21**, 521–537 (2014)
10. Daley, R.: Atmospheric Data Analysis. Cambridge University Press, Cambridge (1991)
11. Dee, D.P., Da Silva, A.: Data assimilation in the presence of forecast bias. Q. J. R. Meteor. Soc. **117**, 269–295 (1998)
12. Dee, D.P., et al.: The era-interim reanalysis: configuration and performance of the data assimilation system. Q. J. R. Meteor. Soc. **137**, 553–597 (2011)
13. Le Dimet, F.X., Talagrand, O.: Variational algorithms for analysis and assimilation of meteorological observations: theoretical aspects. Tellus **38A**, 97–110 (1986)
14. Doblas-Reyes, F., García-Serrano, J., Lienert, F., Pintò Biescas, A., Rodrigues, L.R.L.: Seasonal climate predictability and forecasting: status and prospects. WIREs Clim. Change **4**, 245–268 (2013)
15. Evensen, G.: Data Assimilation: The Ensemble Kalman Filter. Springer, New York (2009)
16. Geleyn, J.F.: Interpolation of wind, temperature and humidity values from model levels to the height of measurement. Tellus **40A**, 347–351 (1988)
17. Ghilain, N., Arboleda, A., Gellens-Meulenberghs, F.: Evapotranspiration modelling at large scale using near-real time MSG SEVIRI derived data. Hydrol. Earth Syst. Sci. **15**, 771–786 (2011)
18. Hamdi, R., Van de Vyver, H., Termonia, P.: New cloud and micro-physics parameterization for use in high-resolution dynamical downscaling: application for summer extreme temperature over Belgium. Int. J. Climatol. **32**, 2051–2065 (2012)
19. Li Hong, L., Kalnay, E.: Simultaneous estimation of covariance inflation and observation errors within an ensemble Kalman filter. Q. J. R. Meteor. Soc. **533**, 523–533 (2009)
20. Janjić, T., Cohn, S.E.: Treatment of observation error due to unresolved scales in atmospheric data assimilation. Mon. Weather Rev. **134**, 2900–2915 (2006)
21. Jazwinski, A.H.: Stochastic Processes and Filtering Theory. Academic, New York (1970)
22. Kalman, R.: A new approach to linear filtering and prediction problems. Trans. ASME J. Basic Eng. **82**, 35–45 (1960)
23. Kalnay, E.: Atmospheric Modeling, Data Assimilation, and Predictability. Cambridge University Press, Cambridge (2002)
24. Kondrashov, D., Shprits, Y., Ghil, M., Thorne, R.: Kalman filter technique to estimate relativistic electron lifetimes in the outer radiation belt. J. Geophys. Res. **112**, A10227 (2007)
25. Kondrashov, D., Sun, C., Ghil, M.: Data assimilation for a coupled Ocean-Atmosphere model. Part II: Parameter estimation. Mon. Weather Rev. **136**(12), 5062–5076 (2008)
26. Leith, C.E.: Predictability of Climate. Nature **276**(11), 352–355 (1978)
27. Lewis, J.M., Derber, J.C.: The use of adjoint equations to solve a variational adjustment problem with advective constraint. Tellus **37A**, 309–322 (1985)
28. Lorenz, E.N.: Deterministic non-periodic flow. J. Atmos. Sci. **20**, 130–141 (1963)
29. Lorenz, E.N.: Predictability – a problem partly solved. In: Palmer, T.N. (ed.) Proceedings of ECMWF Seminar on Predictability, vol. I, pp. 1–18. ECMWF, Reading (1996)
30. Mahfouf, J.-F.: Soil analysis at météo-france. part i: Evaluation and perspectives at local scale. Note de Cent. CNRM/GMME **84**, 58 (2007)
31. Mahfouf, J.-F., Bergaoui, K., Draper, C., Bouyssel, F., Taillefer, F., Taseva, L.: A comparison of two off-line soil analysis schemes for assimilation of screen level observations. J. Geophys. Res. **114**, D80105 (2009)
32. Mitchell, L., Carrassi, A.: Accounting for model error due to unresolved scales within ensemble Kalman filtering. Q. J. R. Meteor. Soc. **141**, 1417–1428 (2015)
33. Nicolis, C.: Dynamics of model error: some generic features. J. Atmos. Sci. **60**, 2208–2218 (2003)

34. Nicolis, C.: Dynamics of model error: the role of unresolved scales revisited. J. Atmos. Sci. **61**(1996), 1740–1753 (2004)
35. Nicolis, C., Perdigao, R., Vannitsem, S.: Dynamics of prediction errors under the combined effect of initial condition and model errors. J. Atmos. Sci. **66**, 766–778 (2009)
36. Noilhan, J., Mahfouf, J.-F.: The ISBA land-surface parameterization scheme. Glob. Planet. Chang. **13**, 145–159 (1996)
37. Orescanin, B., Rajkovic, B., Zupanski, M., Zupanski, D.: Soil model parameter estimation with ensemble data assimilation. Atmos. Sci. Lett. **10**, 127–131 (2009)
38. Raanes, P., Carrassi, A., Bertino, L.: Extending the square root method to account for model noise in the ensemble Kalman filter. Mon. Weather. Rev. **143**, 3857–3873 (2015)
39. Ruiz, J.J., Pulido, M., Miyoshi, T.: Estimating model parameters with ensemble-based data assimilation: parameter covariance treatment. J. Meteorol. Soc. Jpn. **91**, 453–469 (2013)
40. Sasaki, Y.: Some basic formalism in numerical variational analysis. Mon. Weather Rev. **98**, 875–883 (1970)
41. Talagrand, O., Courtier, P.: Variational assimilation of meteorological observations with the adjoint vorticity equation. I: theory. Q. J. R. Meteorol. Soc. **113**, 1311–1328 (1987)
42. Tremolet, Y.: Accounting for an imperfect model in 4D-Var. Q. J. R. Meteor. Soc. **132**(621), 2483–2504 (2006)
43. Tremolet, Y.: Model error estimation in 4d-var. Q. J. R. Meteor. Soc. **133**, 1267–1280 (2007)
44. Uboldi, F., Kamachi, M.: Time-space weak-constraint data assimilation for nonlinear models. Tellus **52**, 412–421 (2000)
45. Vannitsem, S., Toth, Z.: Short-term dynamics of model errors. J. Atmos. Sci. **59**, 2594–2604 (2002)
46. Yang, X., Delsole, T.: Using the ensemble Kalman filter to estimate multiplicative model parameters. Tellus **61A**, 601–609 (2009)
47. Weber, R.J.T., Carrassi, A., Doblas-Reyes, F.: Linking the anomaly initialization approach to the mapping paradigm: a proof-of-concept study Mon. Weather Rev. **143**, 4695–4713 (2015)

Decreasing Flow Uncertainty in Bayesian Inverse Problems Through Lagrangian Drifter Control

Damon McDougall and Chris K.R.T. Jones

Abstract Commonplace in oceanography is the collection of ocean drifter positions. Ocean drifters are devices that sit on the surface of the ocean and move with the flow, transmitting their position via GPS to stations on land. Using drifter data, it is possible to obtain a posterior on the underlying flow. This problem, however, is highly underdetermined. Through controlling an ocean drifter, we attempt to improve our knowledge of the underlying flow. We do this by instructing the drifter to explore parts of the flow currently uncharted, thereby obtaining fresh observations. The efficacy of a control is determined by its effect on the variance of the posterior distribution. A smaller variance is interpreted as a better understanding of the flow. We show that a systematic reduction in variance can be achieved by utilising controls that allow the drifter to navigate new or interesting flow structures, a good example of which are eddies.

Keywords Bayesian inverse problem • Lagrangian data assimilation • Ocean drifter • Control

1 Introduction

The context of the problem we address in this paper is that of reconstructing a flow field from Lagrangian observations. This is an identical twin experiment in which a true flow field is unknown but from which Lagrangian type observations are extracted. It is assumed that little is known about the functional form of the flow field except that is barotropic, incompressible and either steady or with simple known time dependence. Note that, since it is incompressible and two-dimensional (barotropic), the field is given by a stream function $\psi(x, y, t)$. The objective is then

D. McDougall (✉)
Institute for Computational Engineering and Sciences, University of Texas at Austin, Austin, TX, USA
e-mail: damon@ices.utexas.edu

C.K.R.T. Jones
Mathematics Department, University of North Carolina at Chapel Hill, Chapel Hill, NC, USA
e-mail: ckrtj@email.unc.edu

© Springer International Publishing Switzerland 2016
F. Ancona et al. (eds.), *Mathematical Paradigms of Climate Science*, Springer
INdAM Series 15, DOI 10.1007/978-3-319-39092-5_10

to reconstruct an estimate of this stream function from Lagrangian observations, along with an associated uncertainty.

The question addressed is whether the uncertainty of the reconstruction can be reduced by strategic observations using Lagrangian type instruments. The measuring devices are assumed to be controllable and their position can be registered at appropriate time intervals. Since the control is known, being prescribed by the operator, it is reasonable to believe that information can be garnered from the position observations. The issue is whether we can improve the information content in these observations by controlling the instruments to move into specific flow regimes.

Estimating ocean flows has a long history. First, a comparison of model forecast errors in a barotropic open ocean model can be found in [1], with emphasis on how forecasts are sensitive to boundary information. Application of the Kalman filter with Lagrangian observations can be seen as early as 1982 [2–5]. For a variational least-squares approach to eddy estimation, the reader is directed to [6]. A standard mathematical framework for incorporating Lagrangian observations appeared in 2003 [7]. Finally, [8] exposes a novel approach to ocean current observations involving the treatment of sea turtles as Lagrangian observers. A good overview of some operational ocean apparatus can be found in [9].

The underlying philosophy of our approach is that mesoscale ocean flow fields are dominated by coherent features, such as jets and eddies. If the instrument can be controlled to move into and through these structures then the information gained should be richer in terms of capturing the key properties of the flow field.

Of course, there is some circularity inherent in this approach; we want to get the key features of the flow field but need to know them in order to control the vehicle toward them. We first take a "proof of concept" approach and see that following a simple strategy we know takes the glider into another eddy, as opposed to one that does not, reduces flow uncertainty. We then postulate a good way of developing a control based purely on local information. The idea is to use the known local information of the flow field, from reconstructing the flow using observations up to a certain point in time, to form a control that takes the instrument away from the eddy it is currently stuck in.

To obtain the flow reconstruction, one needs to solve a Bayesian inverse problem [10]. There are numerous ways to solve Bayesian inverse problems, with the core methods being Kalman filtering and smoothing [11–16]; variational methods [17–25]; particle filtering [26, 27]; and sampling methods [28–45]. The resulting solution to a Bayesian inverse problem is a probability distribution, called the posterior distribution, over some quantity of interest from which one can compute estimates with associated uncertainties. Bayesian inverse problems enable well-informed predictions.

The paper is organised in the following manner. The second section sets up the Bayesian inverse problem and specifies all the assumptions in the prior and likelihood distributions. The third section applies a proof-of-concept (naïve) zonal (East-West) control of the form $f(x, y) = (\zeta, 0)$ to explore the effect of control strategy on flow uncertainty. This is applied to both the perturbed and unperturbed flows. We measure performance of the control by looking at the posterior variance on the velocity field and show two main results. When the fluid flow drifter is trapped in a recirculation regime, the magnitude of the control is the main player in pushing the drifter out of the eddy. We show that, for the unperturbed flow, when the control magnitude is large enough a significant reduction in the posterior variance is achieved. In the perturbed flow, we show robustness of the posterior variance with respect to the perturbation parameter. More specifically, its structure as a function of control magnitude is carried over from the time-independent flow model. Moreover, we observe an additional, and separate, decrease in posterior variance as a function of control magnitude corresponding to the purely time-dependent part of the flow. The fourth section examines the use of an a posteriori control, a control calculated using information from a previous Bayesian inversion done with no control present. Here the control magnitude corresponds to the distance between the drifter and a hyperbolic fixed point of an eddy transport barrier in the flow. As the control magnitude increases, the drifter gets closer to a hyperbolic fixed point of the drifter evolution equation and, for the unperturbed flow, a substantial decrease in posterior variance is observed. Hyperbolic fixed points of the drifter equations join transport barriers in the flow and act as a boundary to observations. Observing near these points outweighs the negative effects produced by polluting the observations with a large control size relative to the size of the flow. This gives a novel geometric correspondence between the control utilised here and the structure of the posterior variance as a function of control magnitude. The fifth section concludes the paper.

2 Setup

We begin by prescribing the stream function of the flow field the drifters will move in. We will call this flow field the 'truth' and later we try to reconstruct it from noisy observations. The truth flow we will use is an explicit solution to the barotropic vorticity equations [46],

$$\psi(x, y, t) = -cy + A \sin(2\pi kx) \sin(2\pi y) + \varepsilon \psi_1(x, y, t),$$

on the two dimensional torus $(x, y) \in \mathbb{T}^2$. The barotropic vorticity equation in a divergence-free inviscid two-dimensional flow states that the material time-derivative of vorticity is zero. The perturbation ψ_1 we will use is given by,

$$\psi_1(x, y, t) = \sin(2\pi x - \pi t) \sin(4\pi y).$$

The corresponding flow equation is as follows

$$\frac{\partial v}{\partial t} = \varepsilon \partial_t \nabla^\perp \psi_1, \quad t > 0, \tag{1}$$

where the operator $\nabla^\perp = (-\partial_y, \partial_x)$ maps the stream function to its associated velocity field. We will explore two cases. The first case is when $\varepsilon = 0$ and the underlying flow is steady. The second case is when $\varepsilon \neq 0$ and the time-dependent perturbation smears the underlying flow in the x-direction. Drifters placed in the flow v will obey

$$\frac{dx}{dt} = v(x, t) + f(x, t).$$

The function f is called the *control*, the choice of which requires explicit diction. Initially, we consider a proof-of-concept control to illustrate the effect of drifter control on flow uncertainty. Once we demonstrate that drifter control can improve flow uncertainty, we focus on the case where the underlying flow is unknown, and a Bayesian inverse problem must be solved to construct the control. We call these controls a posteriori controls. A posteriori controls harness information from a previous Bayesian update. Our soup-to-nuts methodology for assessing the efficacy for a specific control strategy is as follows:

1. Solve controlled drifter equations. These are coupled to an underlying flow equation;
2. Perform Bayesian inversion and observe the posterior variance;
3. Modify the control magnitude;
4. Go to 1.

More specifically, drifter dynamics are obtained by solving,

$$\frac{dx}{dt} = v(x, t), \quad 0 < t \leq t_{K/2}, \tag{2}$$

$$\frac{dx}{dt} = v(x, t) + f(x), \quad t_{K/2} < t \leq t_K, \tag{3}$$

where v solves (1). Then observations of the drifter locations x are collected into an observation vector for both the controlled and uncontrolled parts

$$y_k^1 = x(t_k) + \eta_k, \quad \eta_k \sim \mathcal{N}(0, \sigma^2 I_2), \quad k = 1, \ldots, K/2,$$

$$y_k^2 = x(t_k) + \eta_k, \quad \eta_k \sim \mathcal{N}(0, \sigma^2 I_2), \quad k = K/2 + 1, \ldots, K,$$

$$\rightsquigarrow y = (y^1, y^2)$$

$$= \mathcal{G}(u) + \eta, \quad \eta \sim \mathcal{N}(0, \sigma^2 I_{2K}), \tag{4}$$

where u is the initial condition of the model (1). The map $\mathcal{G}$ is called the *forward operator* and maps the object we wish to infer to the space in which observations are taken.

The proof-of-concept control f is independent of y^1 and takes the form $f(x, y) = (\zeta, 0)$, acting only in the x-direction. The a posteriori control we execute is one that forces drifter paths to be transverse to streamlines of the underlying flow. Namely, $f(x) = -\zeta \nabla \mathbb{E}(\psi_0 | y^1)$, where $\psi_0(x) = \psi(x, 0)$. Our aim is to understand the effect of the control magnitude ζ and the resulting drifter path on the posterior distribution over the initial condition of the model $\mathbb{P}(u|y^1, y^2)$.

Encompassing our beliefs about how the initial condition, u, should look into a prior probability measure, μ_0, it is possible to express the posterior distribution in terms of the prior and the data using Bayes's theorem. Bayes's theorem posed in an infinite dimensional space says that the posterior probability measure on u, μ^y, is absolutely continuous with respect to the prior probability measure [47]. Furthermore, the Radon-Nikodym derivative between them is given by the likelihood distribution of the data,

$$\frac{d\mu^y}{d\mu_0}(u) = \frac{1}{Z(y)} \exp\left(\frac{1}{2\sigma^2}\|\mathscr{G}(u) - y\|^2\right),$$

where the operator $\mathscr{G}$ is exactly the forward operator as described in (4) and $Z(y)$ is a normalising constant. We utilise a Gaussian prior measure on the flow initial condition, $\mu_0 \sim \mathscr{N}(0, (-\Delta)^{-\alpha})$, where Δ is the Laplacian differential operator. For our purposes, we choose $\alpha = 3$ so that draws from the prior are almost surely in the Sobolev space $H^1(\mathbb{T}^2)$ [47, 48]. The posterior is a high dimensional non-Gaussian distribution requiring careful probing by use of a suitable numerical method. The reader is referred to [47] for a full and detailed treatment of Bayesian inverse problems on function spaces.

To solve the above Bayesian inverse problem, we use a Markov chain Monte Carlo (MCMC) method. MCMC methods are a class of computational techniques for drawing samples from an unknown target distribution. Throughout this paper, we have chosen to sample the posterior using a random walk Metropolis-Hastings method on function space [28, 48, 49]. Using this approach, one can draw samples from the posterior distribution, obtaining its shape exactly. This is of use when the posterior distribution is not a Gaussian and cannot be uniquely determined by its first and second moments. The application of MCMC methods to solve Bayesian inverse problems is widespread. For examples of their use, see [29–37, 50–53]. The theory above is all done in an infinite dimensional setting. Numerically and operationally, a finite dimensional approximation is made. In the case of the Karhunen-Loève expansion this approximation is done by truncation. A choice must be made in where to truncate, and this choice coincides with a modelling assumption—there are no frequencies of order larger than the truncation wavenumber. If it is feasible that solutions to the inverse problem do in fact admit higher-order frequencies, it is necessary to rethink this assumption. Throughout this paper the data and initial conditions are known and the truncation is chosen to be much larger than necessary to mitigate the effects of poor modelling assumptions. Practically, the true solution to one's problem is unknown. In this scenario, care and diligence are necessary in choosing appropriate prior assumptions.

3   Variance Reduction Using a Proof-of-Concept Control

Flow-independent controls concern the influence of an ocean drifter without using knowledge of the underlying flow. They are constructed in such a way as to be independent of y^1. The zonal proof-of-concept control we have chosen acts in the East-West direction and has zero y-component: $f(x, y) = (\zeta, 0)$. Figure 1a shows the variance of the horizontal component of the flow as a function of control magnitude in the max norm, the L^1 norm and L^2 norm. The horizontal axis denotes the strength of the control. The vertical black dotted line corresponds to a critical value for the magnitude. Values of ζ less than this correspond to controls not strong enough to force the drifter out of the eddy. Conversely, values bigger correspond to controls that push the drifter out of the eddy.

Experiments were done for $\zeta = 0, 0.25, 0.5, \ldots, 2.75, 3$. The case $\zeta = 1.75$ was the first experiment in which we observed the drifter leaving the recirculation regime. The black line shows the maximum value of the variance over the domain $[0, 1] \times [0, 0.5]$. The magenta line and cyan line show the L^2 norm and L^1 norm, respectively. The minimum value of the variance is small enough to be difficult to see on the plot but remains consistently small, so it has been omitted for clarity reasons. There are some notable points to make here. Firstly, above the critical value (where the drifter leaves the eddy) we see that the size of the variance decreases in all of our chosen norms. We have learned more about the flow around the truth by forcing the drifter to cross a transport boundary and enter a new flow regime. Secondly, below the critical region (where the drifter does not leave the eddy) we see an initial increase in the size of the variance. There are many factors at play here. We will try to shed some light on them.

For small ζ, the controlled and uncontrolled paths along which we take observations are close. Their closeness and the size of the variance σ^2 creates a delicate

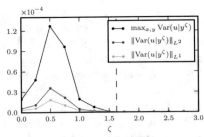

(a) The norm of the variance decreases as the glider is forced across the transport boundary and out of the eddy. The bump occurs as a result of the glider exploring a slow part of the flow.

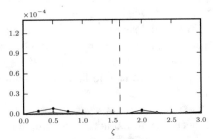

(b) The norm of the variance decreases as the glider is forced across the transport boundary and out of the eddy. The second bump appears because the glider re-enters a time-dependent eddy.

Fig. 1 Posterior variance as a function of control magnitude, ζ, for (**a**) the time-independent model; and (**b**) the time-dependent model on the same y-scale. The *black dotted line* illustrates the max-norm of the variance for the uncontrolled drifter

Fig. 2 True glider path
(*black*) for some positive ζ
less than the critical value.
Blue lines are streamlines of
the true flow. *Red crosses* are
zeros of the flow: fixed points
of the passive glider equation

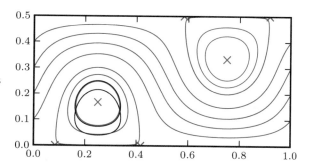

interplay between whether they are statistically indistinguishable or not. If they
are indistinguishable up to two or three standard deviations, this could explain the
increase and then decrease of the variance below the critical value. Secondly, as ζ
increases initially, the controlled path gets pushed down near the elliptic stagnation
point of the flow (see Fig. 2), and here v is smaller in magnitude than v restricted to
the uncontrolled path. Therefore, there is an increase in the magnitude of the control
f relative to the underlying flow v and this leads to the observations becoming
polluted by f.

Exploring this further, we compute the mean magnitude of the flow along the
controlled path of the drifter. More formally, we solve (3) to obtain a set of points
$\{z_k = z(t_k)\}_{k=1}^K$. Then we compute the mean flow magnitude as follows

$$\langle v \rangle = \frac{2}{K} \sum_{k=K/2+1}^{K} v(z_k). \tag{5}$$

This quantity is computed for each fixed ζ and the result is plotted in Fig. 3. The
mean flow magnitude is given by the magenta line in this figure and the black dotted
line depicts the flow magnitude. Notice the first three values of ζ for which the mean
flow magnitude decreases. This is equivalent to an increase in the magnitude of the
control relative to the magnitude of the underlying flow and so the information gain
from taking observations here decreases. This corresponds nicely with the first three
values of ζ in Fig. 1a that show an increase in variance. Also notice that for the other
values of ζ the mean flow magnitude shows a mostly increasing trend consistent with
a decrease in the posterior variance.

Note that the region below the critical value corresponds to control magnitudes
that are too small to push the glider out of the eddy *in the unperturbed case $\varepsilon = 0$*.
The region above the critical value corresponds to values of ζ for which the glider
leaves the eddy. This is also in the unperturbed case. Experiments were done for
$\zeta = 0, 0.25, 0.5, \ldots, 2.75, 3$. In the case $\varepsilon = 0$ (see Fig. 1a), the value $\zeta = 1.75$
was the first experiment in which we observed the glider leaving the recirculation
regime. The black line shows the maximum value of the variance over the domain
$[0, 1] \times [0, 0.5]$. The magenta line and cyan line show the L^2 norm and L^1 norm,

Fig. 3 Mean magnitude of the flow along the control path (*purple*) against the size of the control (*black dashed line*). When the gradient of the flow magnitude is large compared with that of the control magnitude, the posterior variance is small

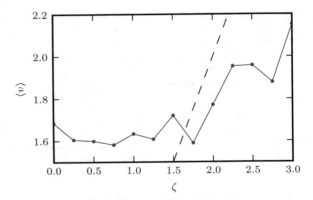

respectively. There are some notable points to make. Firstly, below the critical magnitude (where the glider leaves the eddy in the unperturbed case) we see a sizeable reduction of posterior variance in the max norm as the critical magnitude is approached. Note that as ζ increases in the perturbed case (Fig. 1b) and progresses further into the region above the critical magnitude, the posterior variance repeats the increasing/decreasing structure induced by the eddy that we observed in the region below the critical control magnitude. The new effects introduced into this region are purely from the time-dependent nature of the moving eddy. The reason for their presence is much the same as in the time-independent case; observations trapped within an eddy regime.

We have learned more about the flow around the truth by forcing the glider into the meandering jet flow regime. The benefits of such a control occur at exactly the same place as in the time-independent case; as the drifter leaves the eddy in the unperturbed flow. However, extra care is required when the flow is time-dependent and the eddy moves. One cannot simply apply the same control techniques as is evidenced by the extra bump in variance in the region above the critical magnitude. Of particular use would be extra eddy-tracking information to construct an a posteriori control to keep the variance small.

4 Posterior Informed Control

In Sect. 3 we concluded that crossing a transport boundary and entering a new flow regime has the desirable effect of reducing the posterior variance. Crossing into new flow regimes with a stationary flow can be translated to travelling transversely against the streamlines of the underlying flow. For the recirculation regime of Fig. 2 located in the bottom-left area, particles in the fluid will move in an clockwise fashion. The gradient of the stream function will therefore point in towards the fixed point at $z = (1/4, 1/6)$. The negative gradient of the stream function points towards the fixed point at $z = (3/4, 1/3)$. Therefore, to escape the recirculation regime we

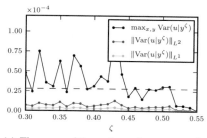
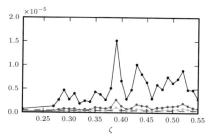

(a) The norm of the variance decreases as the glider is forced towards the a saddle point in the flow. No clear gain is made otherwise.

(b) No clear gain is made in the case of the time-dependent model.

Fig. 4 Posterior variance as a function of control magnitude, ζ, for the a posteriori control in the case of: (**a**) a time-independent model; and (**b**) a time-dependent model

choose,

$$f(z) = -\zeta \nabla_z (\mathbb{E}(\psi|y^1)), \tag{6}$$

for the controlled drifter model, where ψ is the stream function of the flow v. The rationale behind this choice is that, if the posterior mean stream function is a good estimator of the flow, the drifter will be forced transversely with the stream lines and escape the recirculation regime and allow us to make observations in a new flow regime.

Figure 4a depicts the variance of the horizontal component as the strength of the control, ζ, is varied. Note that we do not see the same behaviour as we do for the two naïve controls chosen in Sect. 3. We see a large band of values of ζ for which the posterior variance oscillates, leading to a lack of information gain in the knowledge of the flow. From about $\zeta = 0.5$ to $\zeta = 0.55$, we see a structurally significant reduction in posterior variance where we have a sustained gain in information about the underlying flow field. This is attributed to a drifter path that explores an 'interesting' part of the flow where a lot of information can be obtained from observations. To explore the geometric correspondence between the variance reduction for $\zeta = 0.5$ to $\zeta = 0.55$, we consider Fig. 5. This figure presents the true path of the drifter for $\zeta = 0.3, \ldots, 0.55$. The light pink path corresponds to a value of $\zeta = 0.3$ and the purple path corresponds to $\zeta = 0.55$. Notice that as ζ increases, the true path forms a trajectory close to the zero of the flow at $(x, y) = (7/12, 1/2)$. Just as we have seen in Sect. 3, we observe a transient period in the posterior variance until we utilise a control for which the true path explores new aspects of the flow compared with other 'nearby' controls. Interestingly, also note that we observe this reduction in variance despite the true path navigating near a zero of the flow, where we also satisfy the fact the size of the control is large in comparison to the flow. In this case, a logical conclusion here would be that the information gain from observing near an interesting flow structure

Fig. 5 The true drifter paths
for each value of ζ for the
experiments shown in Fig. 4a.
The *pink path* corresponds to
the magntidue $\zeta = 0.3$ and
the *purple path* corresponds
to $\zeta = 0.55$. The posterior
variance decays as ζ
approaches 0.55

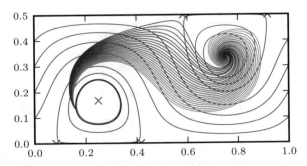

heavily outweighs the information loss in polluting the observations with such a
control. The cost of polluting the observed data can be seen by computing the most
structurally significant reduction in the posterior variance and comparing this with
Fig. 1a, for example. By 'most structurally significant' we loosely mean the most
dramatic reduction that leads to the most benefit in knowledge of the underlying
flow. In this example, this occurs between $\zeta = 0.52$ and $\zeta = 0.55$, where it is
approximately 3×10^{-5}. In the case of the zonal control, where the relative size
of the flow *increases* for the values of ζ that give a reduction in variance, it occurs
between $\zeta = 0.5$ and $\zeta = 1.5$ where it is approximately 1.2×10^{-4}. This is about an
order of magnitude bigger, crystallising the tradeoff between polluting the observed
data versus exploring 'interesting' parts of the flow. If the posterior mean is a good
estimator of the underlying flow, utilising a control of this nature is beneficial if
the drifters navigates close to a hyperbolic fixed point of the passive drifter model
equation.

For the time-dependent case, we turn our attention to Fig. 4b. The first thing to
note is that we do not see the same behaviour as we do for the two naïve controls
chosen in Sect. 3. Nor do we see similar structures when compared with Fig. 4a.
For each value of ζ, it is the case that the true path navigates to the time-dependent
eddy surrounding the zero of the flow at the point $(x, y) = (3/4, 1/3)$. The second
thing to note is that for all of these values of control magnitude, the smaller values
tend to do better than the larger ones. The variance is lower in the cases $\zeta = 0.21$
and $\zeta = 0.27$ because the true path is navigating towards one of the hyperbolic fixed
points of the eddy. A novel connection is established between the behaviour of these
two controls in both the time-independent case and the time-periodic case.

5 Conclusion

To summarise, we have measured the performance of two control methods, a
proof-of-concept control and an a posteriori control, both in a time-independent
and time-dependent flow. We have done so by observing their influence on the
posterior variance in the mean flow direction. Section 3 addresses the proof-of-
concept control and Sect. 4 the a posteriori control. Each control is designed to

push ocean drifters into uncharted flow regimes. The two cases of control we employ here are a purely zonal control, and the gradient of the posterior mean constructed using a posteriori information from a previous Bayesian update. In the time-independent flow, we show a sizeable reduction of the posterior variance in the mean flow direction for these two cases of control. In the case of the a posteriori control in the time-independent flow, the drifter leaves the eddy for all the values of control magnitude we have chosen. Here we observe the variance reduction occurring when the true drifter path approaches a hyperbolic fixed point on the transport barrier of the eddy in the upper-right of the domain. This is evidence that oceanic transport barriers heavily influence posterior information and sets up a novel geometric correspondence between the flow structure and the posterior variance. Using the proof-of-concept control in the time-dependent flow, we show *robustness* of posterior variance as a function of the perturbation parameter. When the control magnitude is such that the drifter leaves the eddy in the *unperturbed* flow, we see reduction in the posterior variance on the initial condition for the time-periodic flow. When employing a time-dependent a posteriori control, we see no overall net gain in posterior variance over the uncontrolled case. For our particular flow and drifter initial condition, it is the case that the uncontrolled drifter path explores a hyperbolic fixed point of an eddy in the time-dependent flow more effectively than the controlled path. This reiterates the efficacy of control strategies and their influence on the path along which observations are made.

There are a number of ways in which this work could be generalised in order to obtain a deeper understanding of the effects controlled ocean drifters have on flow uncertainty. For example, (i) the study of non-periodic model dynamics; (ii) the use of information from the posterior *variance*; (iii) more elaborate control strategies. Many other generalisations are also possible. Non-periodic models are more dynamically consistent with regards to their approximation of larger ocean models. We have seen the application of posterior knowledge in the construction of a control, though only through use of the mean. The variance of the underlying flow could be used in a similar fashion, perhaps to control ocean drifters towards an area of large variance. This could have a similar effect on the posterior distribution as the method of controlling a drifter into a new, unexplored flow regime. Moreover, controls could be constructed to better reflect reality. Ocean gliders have a limited amount of battery power. Utilising this knowledge in designing a mission plan to optimise a glider's lifespan certainly has its practical applications. Controls that minimise the pollution of the observed data is also desirable. Throughout this paper, we have only used information from one previous Bayesian update. Constructing and executing a posteriori control strategies is a paradigm well suited to that of a Kalman or particle filter; updating the control every time an analysis step is performed. This is left for future discussion.

Acknowledgements Author McDougall would like to acknowledge the work of John Hunter (1968–2012), who led the development of an open-source and freely available plotting library, matplotlib, capable of producing publication-quality graphics [54]. All the figures in this publication were produced with matplotlib.

References

1. Robinson, A.R., Haidvogel, D.B.: Dynamical forceast experiments with a baroptropic open ocean model. J. Phys. Oceanogr. **10**, 1928 (1981)
2. Barbieri, R.W., Schopf, P.S.: Oceanographic applications of the Kalman filter. NASA Technical Memorandum 83993 (1982)
3. Miller, R.N.: Toward the application of the Kalman filter to regional open ocean modeling. J. Phys. Oceanogr. **16**, 72 (1986)
4. Parrish, D.F., Cohn, S.E.: A Kalman filter for a two-dimensional shallow-water model. In: 7th Conference on Numerical Weather Prediction, Montreal, pp. 1–8 (1985)
5. Carter, E.F.: Assimilation of Lagrangian data into a numerical model. Dyn. Atmos. Oceans **13**(3–4), 335 (1989)
6. Robinson, A.R., Leslie, W.G.: Estimation and prediction of oceanic eddy fields. Prog. Oceanogr. **14**, 485 (1985)
7. Kuznetsov, L., Ide, K., Jones, C.K.R.T.: A method for assimilation of Lagrangian data. Mon. Weather Rev. **131**, 2247 (2003)
8. Robel, A.A., Susan Lozier, M., Gary, S.F., Shillinger, G.L., Bailey, H., Bograd, S.J.: Projecting uncertainty onto marine megafauna trajectories. Deep Sea Res. Part I Oceanogr. Res. Pap. **58**(12), 915 (2011). doi:10.1016/j.dsr.2011.06.009
9. Rudnick, D.L., Davis, R.E., Eriksen, C.C., Fratantoni, D.M., Perry, M.J.: Underwater gliders for ocean research. Mar. Technol. Soc. J. **38**(2), 73 (2004). doi:10.4031/002533204787522703
10. Kalnay, E.: Atmospheric Modeling, Data Assimilation and Predictability. Cambridge University Press, Cambridge/New York (2002)
11. Kalman, R.E.: A new approach to linear filtering and prediction problems. J. Basic Eng. **82**(Series D), 35 (1960)
12. Kalman, R.E., Bucy, R.S.: New results in linear filtering and prediction theory. J. Basic Eng. **83**, 95 (1961)
13. Sorenson, H.W.: Kalman Filtering: Theory and Application. IEEE, New York (1960)
14. Evensen, G.: Data Assimilation: The Ensemble Kalman Filter. Springer, Berlin/New York (2006)
15. Houtekamer, P.L., Mitchell, H.L.: Data assimilation using an ensemble Kalman filter technique. Mon. Weather Rev. **126**, 796 (1998)
16. Anderson, J.L.: A local least squares framework for ensemble filtering. Mon. Weather Rev. **131**(4), 634 (2003). doi:10.1175/1520-0493(2003)131<0634:ALLSFF>2.0.CO;2
17. Lorenc, A.C., Ballard, S.P., Bell, R.S., Ingleby, N.B., Andrews, P.L.F., Barker, D.M., Bray, J.R., Clayton, A.M., Dalby, T., Li, D., Payne, T.J., Saunders, F.W.: The Met. Office global three-dimensional variational data assimilation scheme. Q. J. R. Meteorol. Soc. **126**(570), 2991 (2000). doi:10.1002/qj.49712657002
18. Bengtsson, L.: 4-dimensional assimilation of meteorological observations. World Meteorological Organization (1975)
19. Lewis, J.M., Derber, J.C.: The use of adjoint equations to solve a variational adjustment problem with advective constraints. Tellus A **37**(4), 309–322 (1985)
20. Lorenc, A.C.: Analysis methods for numerical weather prediction. Q. J. R. Meteorol. Soc. **112**, 1177 (1986)
21. Dimet, F.X.L., Talagrand, O.: Variational algorithms for analysis and assimilation of meteorological observations: theoretical aspects. Tellus A **38**(2), 97 (1986)

22. Talagrand, O., Courtier, P.: Variational assimilation of meteorological observations with the adjoint vorticity equation. Q. J. R. Meteorol. Soc. **113**, 1311 (1987)
23. Courtier, P., Thépaut, J.N., Hollingsworth, A.: A strategy for operational implementation of 4D-Var, using an incremental approach. Q. J. R. Meteorol. Soc. **120**(519), 1367 (1994)
24. Lawless, A.S., Gratton, S., Nichols, N.K.: An investigation of incremental 4D-Var using non-tangent linear models. Technical report, 2005
25. Lawless, A.S., Gratton, S., Nichols, N.K.: Approximate iterative methods for variational data assimilation. Technical report, Apr 2005
26. Doucet, A., de Freitas, N., Gordon, N.: Sequential Monte Carlo Methods in Practice. Springer Science & Business Media, New York (2001)
27. Leeuwen, P.J.V.: Nonlinear data assimilation in geosciences: an extremely efficient particle filter. Q. J. R. Meteorol. Soc. **136**(653), 1991 (2010). doi:10.1002/qj.699
28. Cotter, S.L., Roberts, G.O., Stuart, A.M., White, D.: MCMC methods for functions: modifying old algorithms to make them faster. Stat. Sci. **28**(3), 424 (2013)
29. Cotter, S.L., Dashti, M., Robinson, J.C., Stuart, A.M.: Bayesian inverse problems for functions and applications to fluid mechanics. Inverse Probl. **25**(11), 115008 (2009). doi:10.1088/0266-5611/25/11/115008
30. Cotter, S.L., Dashti, M., Stuart, A.M.: Approximation of Bayesian inverse problems for PDEs. SIAM J. Numer. Anal. **48**(1), 322 (2010)
31. Lee, W., McDougall, D., Stuart, A.M.: Kalman filtering and smoothing for linear wave equations with model error. Inverse Probl. **27**(9), 095008 (2011). doi:10.1088/0266-5611/27/9/095008
32. Apte, A., Jones, C.K.R.T., Stuart, A.M., Voss, J.: Data assimilation: mathematical and statistical perspectives. Int. J. Numer. Methods Fluids **56**, 1033 (2008). doi:10.1002/fld
33. Apte, A., Hairer, M., Stuart, A.M., Voss, J.: Sampling the posterior: an approach to non-Gaussian data assimilation. Phys. D Nonlinear Phenom. **230**(1–2), 50 (2007). doi:10.1016/j.physd.2006.06.009
34. Apte, A., Jones, C.K.R.T., Stuart, A.M.: A Bayesian approach to Lagrangian data assimilation. Tellus A **60**(2), 336 (2008). doi:10.1111/j.1600-0870.2007.00295.x
35. Herbei, R., McKeague, I.: Hybrid samplers for ill-posed inverse problems. Scand. J. Stat. **36**(4), 839 (2009)
36. Kaipio, J.P., Kolehmainen, V., Somersalo, E., Vauhkonen, M.: Statistical inversion and Monte Carlo sampling methods in electrical impedance tomography. Inverse Probl. **16**(5), 1487 (2000)
37. Mosegaard, K., Tarantola, A.: Monte Carlo sampling of solutions to inverse problems. J. Geophys. Res. **100**(B7), 12431 (1995). doi:10.1029/94JB03097
38. Roberts, G.O.: Weak convergence and optimal scaling of random walk Metropolis algorithms. Ann. Appl. Probab. **7**(1), 110 (1997)
39. Roberts, G.O., Rosenthal, J.S.: Optimal scaling of discrete approximations to Langevin diffusions. J. R. Stat. Soc. Ser. B (Stat. Methodol.) **60**(1), 255 (1998). doi:10.1111/1467-9868.00123
40. Roberts, G.O., Rosenthal, J.S.: Optimal scaling for various Metropolis-Hastings algorithms. Stat. Sci. **16**(4), 351 (2001). doi:10.1214/ss/1015346320
41. Beskos, A., Roberts, G.O., Stuart, A.M.: Optimal scalings for local Metropolis-Hastings chains on nonproduct targets in high dimensions. Ann. Appl. Probab. **19**(3), 863 (2009). doi:10.1214/08-AAP563
42. Metropolis, N., Rosenbluth, A.W., Rosenbluth, M.N., Teller, A.H., Teller, E.: Equation of state calculations by fast computing machines. J. Chem. Phys. **21**(6), 1087 (1953). doi:10.1063/1.1699114
43. Hastings, W.K.: Monte Carlo sampling methods using Markov chains and their applications. Biometrika **57**(1), 97 (1970). doi:10.1093/biomet/57.1.97
44. Atchadé, Y.F., Rosenthal, J.S.: On adaptive Markov chain Monte Carlo algorithms. Bernoulli **11**(5), 815 (2005). doi:10.3150/bj/1130077595

45. Atchadé, Y.F.: An adaptive version for the Metropolis adjusted Langevin algorithm with a truncated drift. Methodol. Comput. Appl. Probab. **8**(2), 235 (2006). doi:10.1007/s11009-006-8550-0
46. Pierrehumbert, R.T.: Chaotic mixing of tracer and vorticity by modulated traveling Rossby waves. Geophys. Astrophys. Fluid Dyn. **58**, 285 (1991)
47. Stuart, A.M.: Inverse problems: a Bayesian perspective. Acta Numer. **19**, 451 (2010)
48. Bogachev, V.I.: Gaussian Measures. American Mathematical Society, Providence (1998)
49. Cotter, S.L., Dashti, M., Stuart, A.M.: Variational data assimilation using targetted random walks. Int. J. Numer. Methods Fluids **68**, 403 (2011). doi:10.1002/fld
50. Herbei, R., McKeague, I.W., Speer, K.G.: Gyres and jets: inversion of tracer data for ocean circulation structure. J. Phys. Oceanogr. **38**(6), 1180 (2008). doi:10.1175/2007JPO3835.1
51. McKeague, I.W., Nicholls, G., Speer, K., Herbei, R.: Statistical inversion of South Atlantic circulation in an abyssal neutral density layer. J. Mar. Res. **63**(4), 683 (2005). doi:10.1357/0022240054663240
52. Michalak, A.M.: A method for enforcing parameter nonnegativity in Bayesian inverse problems with an application to contaminant source identification. Water Resour. Res. **39**(2), 1 (2003). doi:10.1029/2002WR001480
53. Kaipio, J., Somersalo, E.: Statistical inverse problems: discretization, model reduction and inverse crimes. J. Comput. Appl. Math. **198**(2), 493 (2007). doi:10.1016/j.cam.2005.09.027
54. Hunter, J.D.: A 2D graphics environment. Comput. Sci. Eng. **9**(3), 90 (2007)

Printed in the United States
By Bookmasters